HIGH TENSION

LIFE ON THE SHANNON SCHEME

To ordinary people
and their extraordinary stories

For Lorna,

MICHAEL McCARTHY

HIGH TENSION

LIFE ON THE SHANNON SCHEME

THE LILLIPUT PRESS • DUBLIN

First published 2004 by
THE LILLIPUT PRESS LTD
62-63 Sitric Road, Arbour Hill, Dublin 7, Ireland
www.lilliputpress.ie

A CIP record for this title is available from The British Library.

ISBN 1 84351 061 8

Set in 11 on 14.5 point Adobe Garamond.
Design by Anú Design, Tara
Printed by βetaprint, Dublin

Contents

Foreword

In 2000 Siemens celebrated its 75th anniversary by signing a unique Archival & Heritage Co-Operation Agreement with ESB. In 2002, as ESB celebrated its 75th anniversary, both organizations collaborated in the publication of a book edited by Dr Andy Bielenberg. It was called *The Shannon Scheme – An Inspirational Milestone.* This year — 2004 — marks the 75th anniversary of the commissioning of the Power Station in Ardnacrusha.

As this book is dedicated to the staff in both our organizations, and to those who contributed to the creation and implementation of the Shannon Scheme, 2004 represents an appropriate opportunity for such a publication.

The Shannon Scheme is an important milestone in the shared history of ESB and Siemens. It was important for the Irish nation and indeed has been acclaimed nationally and internationally as a major engineering and social achievement.

Many earlier publications on this remarkable venture have focussed for the most part on explaining the engineering achievement of what at one stage came to be known as 'The Eighth Wonder of the World'. The social side of the Scheme is every bit as important. Without the Shannon Scheme there would have been no Rural Electrification Scheme and widespread use of electricity might have been confined to our cities and towns for several generations.

This book attempts to tell how the Shannon Scheme impacted on the ordinary lives of people. It had both positive and negative effects but few would dispute that, despite the harsh conditions that those working on the Scheme endured, it brought about significant change that moved society forward.

We are grateful to the author Michael McCarthy, local historian and former television producer, who has completed extensive original research to piece together how the Scheme impacted on life in Ireland. The book tells not only a wonderful story but also provides human insight into how people had to adjust their lives.

The author has uncovered and presented new information and we believe that the publication of this work will contribute significantly to further debate and research of this subject. No doubt this book will provide an even greater appreciation and understanding of the significance of this chapter in our history.

We salute Michael McCarthy for completing and publishing his work. He has not only performed a great service for our respective organizations, but has also done a public service. This is a story that deserves to be told.

This book which builds on our last joint-publication also demonstrates the commitment of both our organizations to safeguarding our shared history and heritage. This, in turn, is an integral part of our corporate social responsibility.

Padraig Mc Manus
Chief Executive
ESB

Lorenz Zimmerman
Managing Director
Siemens Limited

Preface and Acknowledgments

Having lived within earshot of the massive turbines at Ardnacrusha while grow-ing up, little did I think that the shadow of the Power Station would follow me for so long throughout my life. Thirty years ago I began researching the social history of the Shannon Scheme for a number of articles that were subsequently published by Jim Kemmy in *The Old Limerick Journal*. Then followed an hour-long television documentary for RTÉ on the building of the Scheme. After that we parted company for a number of years, except for the odd visit to walk the beautiful banks of the headrace where one could engage quietly with the elegant social language of the architecture and think affectionately and gratefully of those who toiled and even died during its construction. Then in 2002, the 75th anniversary of the ESB, I took part in a collaborative effort to mark the occasion by contributing an article to a commemorative book on the Shannon Scheme edited by Andy Bielenberg of UCC's History Department. By then plans were well advanced for elaborate celebrations to mark the 75th anniversary of the opening of the Shannon Scheme itself in 1929. One of these was Ardnacrusha 75, a weekend seminar held at the University of Limerick organized by Joachim Fischer of the College's German Department, The Hunt Museum, Siemens and the ESB. Listening to the varied speakers, who really brought a freshness to the sub-ject, rekindled in me the old enthusiasm to undertake further research on a topic that was never too far away from my heart anyway. It was at this point that Brendan Delany, Archive & Heritage Manager with the ESB, who has been central to all the celebrations, suggested that I combine the already published material with some extra chapters and have it all in one book to mark this anniversary year.

So followed months of research in the National Archives where Brian Donnelly and his team have done trojan work in streamlining and refining the Shannon Scheme collection. I am indebted to Brian and his colleagues for their generous scholarship and unfailing good humour that got one through many a dusty old file. Most of the photographs are from the marvellous Siemens-ESB collection and here I have to say how I appreciated the expertise and knowledge of Pat Yeats, Gerry Hampson and Martin Cranley of the ESB Archive, for whom long hours, over and above the call of duty, did not seem to matter. I am also grateful to those families associated with the Scheme who loaned me the valuable pictures of their parents and relatives. Thanks, too, to the Military Archive, Cathal Brugha Barracks, Dublin, for the photograph of Michael Brennan, to Bobby Duhig for permission to use his watercolours of the Abbey fishermen, and to Elena Mikailova for allowing me to use her recent paintings of the Shannon Scheme. Four of the chapters are reworked articles from *The Old Limerick Journal.* I am grateful to Larry Walshe, the editor, and to the present editorial committee for permission to use them.

Throughout the past few months Brendan Delany, ESB, has been most helpful in opening doors and in smoothing the occasional wrinkle that arose from time to time. A 'can do' man was how Justin Keating described him at the seminar in Limerick. That he certainly is, and his advice and encouragement have been of tremendous value. Michael O'Connor, of Siemens Ireland, has also been very supportive of the project and allowed me to freely avail of their archival resources. The work, too, of Gerald O'Beirne, *The History of Siemens Ireland,* has been a valuable guide throughout. Both companies are financing the publication, and I would like to think of it as a tribute to their remarkable achievements in this country over the past seventy-five years.

It is only natural with anniversaries that we spend a lot of the time looking back at the event itself being celebrated. But the event is only part of the story. It is also about people. Some of those who built the Shannon Scheme worked at Ardnacrusha for years afterwards and their belonging to the ESB was accompanied by a genuine proprietorial pride in what they had achieved. Johnny Cusack, who worked as a messenger boy with Siemens during the construction period and with the ESB in the Power Station afterwards, is the living proof of that. His fond memories of those pioneering days and his affectionate collegiality for his German and Irish workers are not just a pleasure to listen to, but they are also a testimony to the greatness of those who have gone before. Their sons and daughters took their place and, in turn, they carried on that warm sense of pride in their collective history. Happily, some of them still work there today and so the seamless tradition continues. Recently I was honoured by ESB's Group Hydro-Manager, Pat Naughton, to speak at a function in Limerick for the staff of Ardnacrusha and their families. One could not but be impressed by the sense of history shared by the various tables as those retired along with current staff traversed the decades exchanging reminiscences and memories laced with humour and of course, recollections of those who have gone before. Theirs' is a great tradition and there are many who have nurtured it but none more so than Pat Ahearne and Pat Mackey, both of whom retired recently. The former's father was seriously injured during the building of the Scheme and the latter's dad, Mick, the great hurling maestro, was as towering a figure around Ardnacrusha for many years as he had been on the playing field. This anniversary is a celebration of all who have worked there and cared for it down the years, who were always mindful, like those who built it, that this is a legacy for the next generation.

I look on this book as the slightest opening of the door on the personal and family stories associated with the Shannon Scheme and hope that it will enable others to open it further. There are lots of great tales yet to be told, and not just those of Irish people. The Germans were meticulous in the records they kept; their fantastic photographic collection alone at their company base in Munich is testimony to that. Unfortunately, the Russians took Siemens' files during the war, including those relating to the Shannon Scheme. Fortunately, many of them have now been recovered and the story of their recovery is a thriller in itself. Hopefully, in years to come they will provide the basis for the colour and detail of the German side of their years of exile in Clare and Limerick.

Finally, I would just like to thank all who have helped in the production of this book, particularly my wife Gráinne and daughter Niamh for their hours of typing, to all my family for their patience and support, to Mary de Paor for her corrections and advice with script, and to Antony Farrell, publisher and editor of The Lilliput Press, for his guidance and insights.

1 | Did the Germans Really Know?*

Had you been in Limerick in the summer of 1925, you could have spent your Saturday afternoon visiting the Gaiety Cinema, one of ten in the city, to see an American film called *How Could You, Jean?* in five reels. And then you could have gone on to a smoking concert! Or, perhaps, you may have preferred just to stay at home and read *The Clare Champion*. Had you chosen to read the paper, you would have noted that eleven German engineers were already in O'Brien's Bridge studying the course of the existing canal from Limerick. The contract for the Shannon Scheme was signed in August 1925, and this was only July. The report also had it that a mill had been taken over that would accommodate 500 men. Borings were being taken, and the paper forecast that the village of O'Brien's Bridge was set to become a very important place indeed.

The *Champion* also reported that Major Michael Brennan, GOC of the Southern Command of the National Army, had recently visited the area and speculated that the army might build a military barracks there. Michael, from just over the road in Meelick, had been O/C of the East Clare Brigade and the Flying Column during the Anglo-Irish war a few short years previously. He knew this countryside

---

*Text of a lecture delivered during Ardnacrusha 75, a seminar held in the University of Limerick in April 2004 marking the 75th anniversary of the Shannon Scheme.

■ Major Michael Brennan, GOC Southern Command 1925, later Chief of Staff of the Defence Forces.

like the back of his hand, having traversed it repeatedly during the war as hunter and hunted. It was in this area that he hid the English prisoner General Lucas, moving him from safe house to safe house, ending up in Hartigans' of Doonass.

For him, this visit must have brought back those memories of five years previously when the mood was heroic, patriotic and noble. Sadly these memories were now reflected through the prism of the most recent war, the Civil War, which had brought out everything that was mean and base in the Irish psyche. Just two short years before, the fighting in that war stopped officially. During its eleven-month duration there was brother against brother and neighbour against neighbour. Vicious atrocity and bloody cruelty by the Republicans was met with equally violent counter-atrocity by the Free State Army, as each side attempted to impose its will on the other. Basically, the fight was all about legitimacy. The Republicans claimed theirs' from the first Dáil of 1919, while the Provisional Government said that the people had spoken through the ballot box and their decision would be defended by force, if needs be. Forty per cent of the population opposed the Treaty that was passed in the Dáil by only seven votes. But consensus had fractured and so was sidelined. Terror, torture and reprisal took its place. The murder of five Free State soldiers in Knocknagoshel, by a booby-trapped landmine, was met by the counter-murder of Ballyseedy when nine anti-treatyites were strapped to a mine and blown up, killing eight of them. In desperation, the government resorted to what has been called institutionalized killing – when the State actually murders its own under the guise of law – by executing seventy-seven Republicans. This number would have been much greater if others in the Cabinet had had their way. Some commentators put the total killed in the Civil War as high as four thousand. The government admitted to losses of eight hundred.

■ Ardnacrusha shortly after the official opening in 1929.

After eleven months of conflict, both sides collapsed in exhaustion, having lost many of Ireland's brightest and best leaders. The infrastructure of the country was in tatters. Roads had been trenched, railway lines torn up, bridges downed and property destroyed. The bill for prosecuting the war came to £17 million, while the cost of damage to property and social infrastructure was put at £30 million. The total cost was nearly twice the annual budget, putting the country on the brink of bankruptcy, prevented only by a massive loan from the Bank of Ireland.

An idea of relative costs can be had when one considers that in 1925 the contract or the rebuilding of the front block of the GPO, damaged in 1916, was awarded to Alexander Hull and Co., Ringsend, who had tendered at £50,000.

At any rate, as Paddy McGilligan, the Minister for Industry and Commerce, and Carl Freidrich von Siemens worked out the final details of the contract for the Shannon Scheme, the government faced an enormous crisis. The 55,000-strong Irish Army, on which it so depended for its survival, threatened to mutiny. This was coming for a while as resentments built up within the various ranks. Some objected to the presence of so many former British army officers, while

■ Paddy McGilligan, Minister for Industry and Commerce.

others resented an IRB cell that had surfaced and allegedly interfered with promotions. Then a group of Republicans within the army issued an ultimatum to the government to declare itself for the ideal of a Republic. These factors, coupled with the government's announcement that it was going to reduce the army by 40,000 men, triggered the dangerous stand-off. Young Kevin O'Higgins, the Vice-President of the Executive Council (Tánaiste/Cabinet) and Minister for Home Affairs (later Justice), handled the crisis most adroitly. Reportedly, he sent Cosgrave home to bed, and then successfully out-manoeuvered his colleague in government, the Labour Minister Joe McGrath, who played a curious role in the affair and whose sympathies would have been with the Republicans within the army. As a result, McGrath resigned and was subsequently appointed Labour Advisor to Siemens. He alleged bitterly that a clique dominated the government. The Minister for Defence, Richard Mulcahy, also was forced to resign. Eight back-benchers followed suit, as did a variety of major generals, colonels, commandants, captains and lieutenants, the whole affair culminating in a purge of the army of Republicans that went right on into 1925 and afterwards. Fortunately, though, civil law had asserted itself over the military – but still these were volatile, dangerous times.

In spite of this and, indeed, strapped for cash, the government determined to push ahead with its plan for demobilization. During the War of Independence it used to cost the British government £210,000 per week to maintain its 55,000-strong army. This gives us some idea of what the National Army was costing the Irish government and why it was so anxious to reduce it. It also had 12,500 Republican prisoners to maintain and was keen to decamp these too, but was

■ Carl Freidrich von Siemens (1872–1941).

hampered from doing so by security considerations.

So, 40,000 men made their way home from their barracks to their villages and towns, where there was little work and much hardship at the time. There they had to live cheek-by-jowl with the neighbours, many of whom were Republicans who still had their guns, grenades, bombs and explosives since the Civil War. In their parishes they had to queue at the local creamery with the same lads who had been their most loyal and brave comrades during the War of Independence, upon whom their lives depended; these same lads whom they now despised for taking the other side in the Civil War; these same lads living in fear of each other and between whom trust was fractured and suspended, and would be for decades; these same lads whom they were now expected to tog out with for the parish and the county. Incredibly, they did.

Inevitably, many old scores were settled, some arising from the Civil War, others from agrarian disputes. There was a celebrated case in Sixmilebridge of the chap who just couldn't bring himself to shoot his neighbour, so he went into his field and with his revolver shot his heifer instead!

From Sixmilebridge to Cratloe, to Woodcock Hill to Glenlon to Broadford, Claremen were engaging in agrarian outrage, shooting, and intimidation. Marauding bands infested the countryside. Many of these were made up of deserters from both armies, but times were desperate. In an incident in Ardnacrusha (now Parteen) four armed men, with their faces blackened and wearing trench coats, broke into the house of James McMahon and said they were searching for arms. Anyway, they tied up his family and made off with some clothes, 3s. 6d. and all the tea and sugar Mrs McMahon had in the house! Times were very hard indeed. There was no work and men had been coarsened by the gun and war.

The only group who stood between this lawlessness and the government was the unarmed civic guards, recently introduced. These had not as yet been accepted by the people and were abducted, shot at and burned out of their barracks. Known as 'Trucileers' and looked on by Republicans as usurpers of the sovereignty of the first Dáil because they were supplanting the Republican police who had come into being with the Republican Courts, they found themselves in a very difficult position. Chosen by Michael Collins from among the various Flying Columns, these battle-hardened young men were brought to the RDS in Dublin to be trained for the new police force. Then they were transferred to Kildare where they mutinied against the presence of so many ex-RIC officers as trainers. Michael Staines, the Commissioner and TD, was forced to move to Dublin, while Paddy Brennan, TD, Major Brennan's brother, was elected Commissioner. Needless to say, there were lots of Clare lads among the recruits. They took no great part in the Civil War, but now found themselves back in the country policing many of their former brave comrades in the War of Independence who had gone with de Valera in the Civil War. As the old song went, 'A policeman's lot is not a happy one.' Theirs' certainly was not and one can understand the story of one barracks, where a sergeant and his men cleaned the barracks, signed the day book, locked the door, put the key in the postbox and emigrated to America.

That is more or less the position in which the eleven young German engineers found themselves in July 1925. Did they know anything of it? Had they even read *The Clare Champion* headlines they would have got an inkling: 'Rifle Rule in Clare', 'More Revolver Talk', 'Rule of the Gun'. They were dealing with a government that was paranoid about security, a commercial infrastructure that was in bits, and a public that was ambivalent about law and order. Hardly the most auspicious time for a massive international project to get under way! In early 1926 an inter-departmental conference of Finance and Defence considered Major Brennan's report for the protection of the Shannon Scheme. He certainly had been very thorough and in his recommendations had even taken into account similar operations in other countries.

The main points of his report included:

1. The building of a military barracks for 300 men at Ardnacrusha. These would be predominantly infantry, but there would also be some air-force personnel included;

2. A landing strip, as well as a hanger for two planes;

3. Machine guns and anti-aircraft guns on emplacements at the power house.

At O'Brien's Bridge he would accommodate 25 men, with machine and anti-aircraft guns at the weir. Both locations would be restricted areas. The price tag for these developments was put at £41,000. Brennan was leaving nothing to chance and his report is a classic for clarity and detail.

A subsequent report based on police intelligence identified four groups as representing the main threat to the project.

The first consisted of the Irregulars – the Republicans, the IRA, who still had guns, explosives, grenades and mines. But, interestingly enough, they never attacked the undertaking even though it was exceedingly vulnerable throughout the 8.6 miles of operation.

The next group was the Communists. The years leading up to the Shannon Scheme were of widespread industrial unrest. There was a plethora of strikes. It seemed that practically everybody – from domestic servants to carters, to shirt-cutters, to coal workers, to dockers, to agricultural labourers (to whose wage the Shannon Scheme workers wage was pegged) – wanted better conditions and were prepared to strike to get them. Expectations were high and it wasn't just the 'landless men', 'the men of no property' who, with independence, felt they were entering the Promised Land. Workers expected more of 'their own', as it were. The One Big Union of Liberty Hall was on a roll, supplanting many of the smaller British combines. At this stage, Labour wasn't for turning, or for waiting.

Like so many all over the world, thousands of Irish workers borrowed pace from the Bolshevik Revolution of 1917. Ireland had bought into the ideals of the French Revolution for 1798, and the Jacobin idea of *saoirse* for an earlier campaign. Similarly, after the Russian Revolution, workers here harnessed the socialist model for development and set up revolutionary workers groups. A Provisional Soviet Government was declared in Carlow; Limerick City developed a major Soviet, another was declared at Cleeve's condensed milk factory in Knocklong, just out the road, while yet another sprang up at Violet Hill, in Broadford, the parish next door to where the Shannon Scheme was being undertaken. Workers' hopes were high and they lustily sang Jim Connell's 'The Red Flag', in spite of the running condemnation of socialism by the Catholic Church. People heard a lot about the cloven hoof of socialism in those days, and for years afterwards. But an interesting feature of the time was the military intelligence reportage from the Limerick/Clare areas of the Southern Command. Aodh de Bláca, Intelligence Officer and rapporteur for the Command, indicated that they were very worried about Irregular guns falling into Bolshevik hands to be used in labour disputes.

The third group identified by the report was 'unemployed malcontents' of whom there were many. It was feared that these would engage in Luddite activity just to slow down the project. Almost 5000 men were employed when building was at its peak. Interestingly, while there was a big controversy in Clare about direct labour for road work and some of those taken on for such work were shot at and intimidated, there is no evidence of similar attacks on the workers at the Shannon Scheme.

The massive strike over labourers' wages at the very beginning of the project left a huge legacy of bitterness and ill will towards the authorities. Joe McGrath, the former Minister who had first introduced the Shannon Scheme to the Dáil and who was now Labour Advisor to Siemens, broke the strike by introducing ex-servicemen whom he could rely on. His strong-arm tactics incensed the unions – he being a former union worker. But that wasn't the first time he had crossed the unions. He had faced them down in the second big postal strike in 1922, when over 10,000 civil servants took to the streets. Picketers were intimidated and shot at. This time, operating out of Limerick's Strand Barracks, he quickly engaged and defied the unions who should have known who they were dealing with. I recall an interview a few years ago with Major General Mickey Jo Costelloe, one-time Director of Intelligence for the army. He was of the opinion that McGrath, 'brought up at the back of the pipes in Dublin,' as he put it, was one of the toughest men he had ever met. He was a colourful, bluff, brawling individual who at one time worked with Michael Collins in the Craig Gardner accountancy firm. They were good friends and, I suppose, it was not surprising that McGrath's brother George, who also worked in Craig Gardner, became the first Comptroller and Auditor General of the Free State. Joe had fought in 1916 and while he had been in and out of gaol, he personally took very few prisoners. His ruthlessness was well known, not just when he was involved with the Sweep, but even at an earlier stage. In 1926 he was embroiled in a renowned libel case against an English author and a publisher of a book that claimed McGrath personally was involved in the murder of British agents when he was Head of the Secret Service. If there was one person who could sort out the seemingly intractable problem on the Shannon Scheme it was McGrath. And he did, but it was a long, tough haul. He left the unions defeated and demoralized.

However, McGrath was not without his critics. As a controversial character, he had many, but it was his employers, Siemens, who complained most loudly about him to the government. Their beef was not about his handling of the strike, but that he, as Director of Labour, was unable to supply what they would consider quality workers for the Scheme. This was no fault of McGrath's. Most Irish workers had never experienced life on a massive building site before. And, while he quickly managed to set up a system of labour recruitment from exchanges throughout the country, the contractors still complained. There is a file in the National Archives dealing with Irish workers. The overriding impression one gets from it is how superior the German workers felt in comparison with their Irish colleagues. At one point the Germans even did a manpower survey to prove how superior they were! Much of the German criticisms were true in that few of the Irish labourers had experience in working with heavy machinery on enormous building sites, so the pace of progress was not what the contractors wanted and the cost in lives and horrific accidents was exceptionally high. Thirty-three men were killed on sites during construction. Dismissals were frequent, sometimes peremptory, and many of those sacked hung around the area in the hope of being taken back on. Some could be classed as 'unemployed malcontents'. There were a few cases of these men causing trouble, like interfering with railway tracks or throwing wire over electric cables, but there was nothing major.

The fourth and last group who, according to the report, could pose a danger to the Scheme was composed of 'foreign agents'. Now, unquestionably, the signing of this contract with a German firm got up imperial noses, not only in Britain, but here also. In 1925 we were still very much part of the empire. Our ministers attended Commonwealth conferences and they got their seals of office from the Governor General, in the Vice-Regal Lodge. Even as late as 1932 when the Electricity (Supply Amendment) Bill, which allowed the government spend an extra £335,000 on the Shannon Scheme, was passed, the Executive Council had to send it to the Governor General, James McNeill, under the formula, 'The Executive Council present Your Excellency for signification in the King's name and for the King's assent, etc., etc.' The ubiquitous royal shadow even followed you to your grave. A medical inquest report form began with the words, 'An inquisition for the King of Saorstát Éireann ...' This used to get on the medical wick of poor Dr Lynch, the coroner in Ennis, as when filling out reports he crossed out the references to the king.

Predictably, *The Irish Times*, then strongly unionist and reflecting the doubts of its constituency about the wisdom of the Scheme, expressed grave concern that the contract had gone to a German firm. This would have been true to form, considering its robust anti-Kaiser policy during The Great War. The paper

■ O'Brien's Bridge and Weir showing the large area now under water, 1930.

was worried about the 'Teutonising' effect on the Free State of such a large colony of Germans! When one reads the Paddy McGilligan papers in UCD, one just can't help being impressed by the amount of preparation and research he did before signing the contract, but it also becomes clear that the Swedes, the Canadians, the Germans and the Swiss were away ahead of the British in the field.

Our own groups with a vested interest in electrification as a private enterprise were well represented by that old unionist war-horse, Sir John Keane from Waterford, who was the first member of either House of the Oireachtas to oppose the Scheme. He wanted to have nothing to do with what he called a 'profligate proceeding'. Sir John spoke for many in financial circles and for big business generally at the time, and had a very jaundiced view of the young Minister for Industry and Commerce who, in Sir John's opinion, had as much experience as Howdy Doody in running the economy or in negotiating international contracts. The very thought of this young man spending five-and-a-half million pounds or one-fifth of the annual budget on what Keane considered to be a dodgy project was too much for him and many like him to comprehend.

The English press, particularly the high-church *Morning Post*, looked on the granting of the contract to Siemens as being mildly treasonable and an unfriendly action against British industry. Elements of that press, and not just the popular press, were paranoid about this country's relationship with Germany. *The Northern Whig* and *The National Review* lambasted Irish politicians for deciding in 1923 not to collect the tax on German imports, as stipulated in the German Reparation Act of 1921. They also accused the Irish of throwing themselves into the arms of German industry with a succession of contracts for iron pipes, lead pipes, cheap housing and electric cables. They even went so far as to claim that one company, Deutsche Kabelwerke, could boast of monopoly status in Ireland. *The National Review* alleged that the Norddeutscher Lloyd had cynically procured the Pope's blessing for its ships in order to attract Irish passengers who wanted to go to the Eucharistic Congress in Chicago in 1926, and saw something extremely sinister in the fact that Michael Brennan spent his holidays in Germany in 1927! It did not occur to these publications that the low price of German production and the low rate of the deutschmark were very attractive propositions for a government with little money.

So, who were the foreign agents the government feared? In October 1926 Kevin O'Higgins claimed in a speech in Dungarvan that certain people in Dublin city and suburbs had been canvassed recently to join or subscribe to a secret organization whose aim was 'to combat certain anti-British trade tendencies'. The canvassing agent was an ex-officer of the army. He didn't say which army. But he went on to say that this group was now concentrating on the Shannon Scheme in order that the project would be aborted. He attributed the current strike over wages to these conspirators and threatened them with the fullest powers of the State if they persisted.

He could very well have been referring to the British Fascist movement that was run on military lines and had a branch in Dublin. Its HQ was in Liverpool and The Free State was its 'command', a term which rankled in certain quarters here. The group was heavily anti-Bolshevik, according to intelligence reports, but had little clout.

Other possible agents, such as elements of the Irregulars, for example, ex-British army men or ex-Black and Tans, might have tried to sabotage the undertaking, but they didn't.

De Valera and the Republicans looked on W.T. Cosgrave's government as an imperial creation, as being in Britain's back pocket, and economically we were; Britain was our main trading partner and was to remain so for decades. But the signing of

■ A German cargo ship arriving in Limerick docks with materials for the Shannon Scheme.

the contract with the Germans may have helped to put some emotional distance between Ireland and Britain and to swing the compass away from London towards Berlin, even for a while.

As mentioned earlier, the cost of the protection works at Ardnacrusha and O'Brien's Bridge was put at £41,000. Provision was made for this in the government estimates of 1926/1927, but Finance was careful not to emphasize it too much. Ernest Blythe was Minister for Finance. He had succeeded in reducing government expenditure from £43 million in 1924 to £27 million in 1926/1927, and he cut the old age pension from 10 shillings a week to 9 shillings a week, an indelicate move he is less than affectionately remembered for! However, he kept the gimlet eye on Defence, as it was felt that the army was using the Shannon Scheme project to bolster their strength. Peadar McMahon, formerly Lieutenant General of the Limerick IV Brigade, was Secretary to Defence and he had a difficult enough job, having had three different Ministers in as many years.

Blythe felt that the plans for the protection of the Shannon Scheme were too grandiose and the financial estimates flaky. Consequently, he put the proposals on a bureaucratic merry-go-round by exerting pressure on the Army's newly formed Corps of Engineers to build the accommodation for the troops. Defence then said it was up to the Board of Works to do so. And then Finance, Defence and the Board of Works together intimated that the contractors should do the work. Needless to say the contractors refused. By 1928 Blythe's pressure began to pay off.

Later that year, the Army came up with a revised plan costing less than half the earlier estimate of £41,000. It was now proposed to erect an electric fence at a radius of 300 yards around the powerhouse. The voltage would be strong enough to kill a man! Just sixty troops, and not the earlier figure of 300, would be accommodated within the wire. Thirteen or fourteen pillboxes, situated around the periphery of the fence, would house machine gunners.

At O'Brien's Bridge the army had bought Parteen Villa, the home of Mr Jeremiah O'Sullivan, for £2000. It was proposed to modify this house and make it suitable for the accommodation of troops, who, like those at Ardnacrusha, would be supplied by the garrison at Limerick.

The Board of Works, Defence and Industry and Commerce passed this new plan, but Finance was still not convinced. As late as 25 January 1929 it was still asking if constructive protective works were necessary and whether the latest plan met the wishes of the government generally.

Three days later a report on security of the Scheme arrived on the Minister's desk from Professor Rishworth, the Chief Civil Engineer, and Dr Sothman, the Chief Electrical Engineer. Both were of the view that protective works were necessary because of the ease of approach and vulnerability of the weir and the Power Station. They recommended a team of inspectors for the headrace on 24-hour supervision. They also proposed telephones and lights at regular intervals along the length of the headrace that should have an armed patrol in times of unrest. At the bottom of the single-page report there is a one-line handwritten note: 'Establishes the need for protective arrangements!' Three months later agreement in principle was reached among all parties but nothing was ever done. Within a few weeks the Scheme was officially opened but the hot potato had not gone away.

Eventually, the issue of protection works at Ardnacrusha and O'Brien's Bridge made it on to the agenda of the Executive Council for 19 May 1930. However, the question being put to the ministers was, who would be responsible for security – the army, the police or the recently formed ESB? Attached to the agenda was a report from General Eoin O'Duffy, Commissioner of the Garda Síochána.

O'Duffy's report was a comprehensive one outlining his proposals for security of the Scheme. He first of all scotched the Army's idea of an electric fence. Out of date, he claimed; he suggested barbwire instead! He then proposed a steel fence across the headrace to stop mines that could be thrown in at Blackwater Bridge from floating towards the intake building. He would have another fence in the tailrace to prevent armoured steel boats attacking from the Limerick direction. The whole area would be restricted, as would the ground around the weir. Staff would be checked out just in case subversives had infiltrated. He would not share command with the army as he had enough policemen with experience of arms. As a final cost-cutting measure, he would only have single men engaged in the operation as this would help reduce expenditure by not having to pay rent allowances! The cost of his plan was put at £7000. Needless to say, he got the job, but not formally until 1933. In the meantime, two sergeants and eleven gardaí, temporarily housed at O'Brien's Bridge, formed the protection party for the Scheme. Mrs O'Curry's house in Ardnacrusha was acquired by the Department of Justice and turned into a barracks for a sergeant and four gardaí.

After all the teething problems of the fledgling state, these men, the gardaí, had come to establish their own legitimacy with the people as guardians of the peace. My own Dad, who, at one time, was a Captain in the IRA in West Cork, was one of those who minded Ardnacrusha.

2 | No Light without Heat

In 1925 the townland of Ballykeelaun, in south-east Clare and about three miles from Limerick City, took on the appearance of an Irish Klondyke. From every corner of the country men converged on the little village of Ardnacrusha in the hope of securing one of the 3000 jobs on the Shannon Scheme. What was once a quiet farming area with a leisurely agricultural pace became overnight a giant building site blasted and torn by explosives and heavy machinery. The people of the area, with its few shops and three pubs, found themselves swamped by thousands of navvies who were 'housed' in nearby huts, stables, henhouses and barns.

The Shannon Electricity Bill became law in June 1925. On 13 August the contract between the Free State government and the German company of Siemens Schuckert was signed. The contract was to cost £5.2 million and was to be completed within three-and-a-half years.[1] Engineers had already arrived from Germany and begun work.[2] Sites were pegged out and routes prepared for the 60 locomotives, 62 miles of railway track, hundreds of trucks, cranes, stone crushers, diggers and other heavy machinery soon to be imported from Hamburg.

One rather curious early achievement of the Germans was to transpose the name Ardnacrusha from what is now known as Parteen to the Power Station site. Parteen, in fact, was farther down the road where the Protestant church is located.[3] Seemingly, the Germans had difficulty in establishing which townland was where,

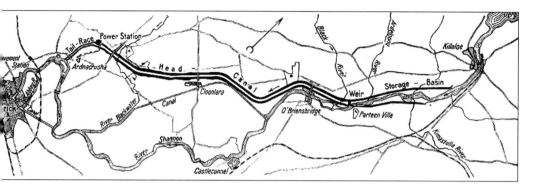

■ An outline of the project stretching from Killaloe to Limerick.

so they called their construction site Ardnacrusha. There are strains of Macbeth and Birnam Wood about it all.

At Longpavement, just outside the city, the Germans set up their railhead from which they ran their network of lines. On Thursday 10 September work was to have started on the building of a line from Longpavement to Ardnacrusha. Labourers on the job were to be paid 8d. an hour for a 54-hour week.[4] The district rate for labourers in Limerick and Clare was 1s. 1d. for a 50-hour week. Labourers who had been taken on by Siemens to convert the Strand Barracks into a storage depot were being paid 1s an hour for a 50-hour week. The city rate was 1s. 3d. per hour for a 47-hour week. Besides, skilled trades on the Siemens' job were being offered less than the district rate.[5] On the first morning of the Longpavement job a group of ex-servicemen of the national army approached the labourers there and urged them to stop work until the recognized rate was paid. The men agreed to do so. The Irish Transport and General Workers' Union (ITGWU), which represented many of the labourers, immediately issued a state-ment saying that work should continue, but that talks on conditions and wages should take place without delay.[6] This advice was ignored by the men. The Transport Union then approached Siemens but, since the contractors would not meet its demands, formal instructions were given by the union for the work to cease.[7]

The stage was now set for what was to be a long, hard, bitter struggle. Organized labour was facing one of its greatest challenges since the foundation of the state, and viewed it as such.[8] There were four things in dispute:

1. the wages of unskilled labour outside the city; 2. the wages of skilled labour outside the city; 3. the wages of unskilled labour in the city, and 4. the hours of the working week in and outside the city.[9]

■ One of sixty locomotives arriving from Germany.

One of the most immediate effects of the dispute was to split the ranks of the ex-servicemen's association. On Sunday 13 September, 600 former members of the national army held a meeting in the Transport Union Hall at 91 O'Connell Street, Limerick. The meeting decided to form a new association because men were dissatisfied with the way the old association of ex-service men was handling their affairs. It also decided to affiliate the new organization to the Transport Union in order to achieve greater clout in industrial action. The meeting was conducted by ex-Brigadier Tim Murphy.[10] There is the probability too, of course, that the association was splitting along Free State versus Republican lines, as the strains of the Civil War were hardly out of earshot. At a further meeting on the following Friday John Mulqueen was elected secretary of the new body, which had adopted the title of ex-national Army Officers and Men's Section, Irish Transport and General Workers' Union.[11]

The old association, led by ex-Captain Terry Casey, protested strongly and publicly in the local newspapers. It objected, firstly, to the formation of the new organization and, secondly, to the affiliation of the ex-servicemen to the Transport Union.[12] Casey maintained that his group represented 90 per cent of

■ German workers were attacked in the docks area throughout Limerick city during the strike.

ex-servicemen in the city and area.[13] However, it was Mulqueen who made most of the running in the early days of the dispute, with letters to the press protesting against the government and low wages and calling on the 800 ex-servicemen in the city not to apply for the job.[14]

On 25 September an advertisement appeared in the national press for 3000 unskilled workers for constructional work on the Shannon river area in counties Limerick and Clare. Wages were set at 32s. a week for a 50-hour week, with free lodgings. The advertisement also stated that canteens were to be established and run by Irish contractors, where cooked food and other necessities could be bought for cost price. Applicants were requested to fill the printed form and address it to 'The Shannon Scheme, Dublin'.[15] Interestingly, the Siemens headquarters in Dublin were at 43 Upper O'Connell Street, at one time the offices of the National Land League, the body that had broken landlordism in Ireland.

There was an immediate outcry against the wages and conditions as advertised in the press. The ITGWU, the Irish Women's Workers Union and the ex-servicemen's association affiliated to the Transport Union were unanimous in their condemnation.[16] *The Manchester Guardian* commented:

> Such a wage is no doubt being paid and received contentedly enough on hole-in-the-corner jobs in many Irish counties, but the Shannon Scheme contractors must have been particularly sanguine if they really hoped that such a wage would be accepted from them when a great national scheme has to be put through under the auspices of the Government and when workers have to be collected from a distance. Labour naturally demands that the Government should compel the contractors to act the part of the model employer.

Dr Thomas McLaughlin, who played such a crucial role in getting the Scheme off the ground and was now Siemens' representative in Ireland, entered the controversy on 28 September. He claimed that the rate being offered compared favourably with the going rates for agricultural labourers, and it was these, not city labourers, who should be used as the proper basis for comparison. The average agricultural wage at the time was 25s. for a 57–60-hour week.[17]

The Voice of Labour was quick to reply that it was unfair to compare the Shannon Scheme labourer to the farm labourer who often received some extra benefits in kind over and above the 32s. a week. However, it blamed Paddy McGilligan, the Minister for Industry and Commerce, for the whole situation and for signing the contract without the advice of the trade union movement. 'It was signed as if the working class had no rights, no authority, no recognized status in the State. The Minister had acted in the spirit of an age before the repeal of the Anti-Combination Acts in 1824.'[18]

On 30 September Joseph McGrath, the former Minister for Industry and Commerce, was appointed as Director of Labour for Siemens Schuckert.[19] His appointment angered and dismayed the unions. McGrath had been an organizer with Jim Larkin's union, the Workers Union of Ireland, in Dublin. He was known as a no nonsense individual when in the union and was prepared to use his fists to get his own way at meetings. As poacher turned gamekeeper he was equally ruthless. At the time of his appointment as Director of Labour for Siemens he had been out of work and in financial difficulties, having resigned from the Dáil some months previously because of the Arms Crisis.

McGrath lost no time in getting into his new job. On the very same day as his appointment he opened negotiations with the Transport Union. No agreement was reached.[20] However, McGrath was determined to finish the dispute one way or another. On Thursday, 1 October, he approached Casey's branch of the ex-servicemen's association, which was unaffiliated to the Transport Union, with an offer of 50s. a week and unbroken time.[21] There was considerable surprise and anger in the city when forty-five of these men accepted the offer, and on the next day began work at the Strand Barracks, the docks and the Longpavement.[22] Pickets were immediately placed on these jobs and on the Siemens' office at O'Connell Street, Limerick.[23]

The collective anger of the strikers boiled over on Friday night. Six labourers who had been working at the Strand Barracks went to the Transport Union office in O'Connell Street to explain their position. They were fortunate to escape with their lives.[24] At the Strand Barracks itself a group of the ITGWU affiliated ex-servicemen attacked the unaffiliated strikebreakers. Gardaí were called and after a few charges the strikers dispersed.[25] German workers were also the targets of strikers' violence and these attacks were condemned from various pulpits throughout the city.[26]

By Saturday morning, the strike had begun to spread. The Limerick Trades' Council issued a recommendation to carpenters and joiners in the Strand Barracks not to work with unskilled labour.[27] Members of the dockers section of the Transport Union objected to German mechanics unloading material from ships guarded by the military with fixed bayonets.[28] A meeting of dockers was arranged for Sunday night, and even at this stage it was evident that the dockers would come out in support of their striking fellow workers. Their support was considered vital for the effectiveness of the strike.

On Sunday, 4 October, came the first test of public opinion in the city. Thousands of people turned out for a mass meeting held at the O'Connell monument at 3 pm. Tom Irwin of the Labour Party Executive, William O'Brien, General Secretary ITGWU, T. Kennedy, Vice-President ITGWU, and J. Cronin, President of the Limerick Trades' Council, addressed the gathering and called for solidarity in resisting this attack on the living standards of workers. Cronin, however, counselled against violence: 'A man who strikes another in the fight is no friend of the labour movement. Violence will not be tolerated.'[29]

The advice of the Trades' Council President was quickly forgotten. On the following day the dockers, who had decided to strike, assaulted the German mechanics who were unloading the Norddeutscher Lloyd steamer *Arabia*, which had arrived in port with 1675 tonnes of material.[30] Guards on duty at the docks also came under attack. Strikers, armed with clubs and rubber pipes filled with lead, again set upon a few Germans in the city.[31]

On Monday there was a replacement of the picket at the Strand Barracks, which had been removed following the fracas on Friday night. A new development was the placing of pickets on various premises in the city which had been supplying materials to the German contractors. In spite of the pickets, Casey and approximately 120 of his men turned up for work, but the skilled workers accepted the call of the Trades' Council and stayed at home.[32]

Hopes were raised briefly on Tuesday, 6 October, when the government met representatives of Siemens in an attempt to find a way out of the deadlock. Nothing came of the meeting.[33] On the same day the National Executive of the

■ Dr Thomas McLaughlin played a central role in the Shannon Scheme and was the first Managing Director of the ESB in 1927.

Labour Party issued a statement denying the pay offer as being unmistakably 'an unchristian wage' and calling upon 'all workers whether organized or at present un-organized, to decline to allow themselves be made instruments of this attack on the working men and to refuse to accept work at those wages'.[34] It was hardly this call of the Labour Party, but more likely the violence of the dockers, which influenced Casey and his men but, at any rate, they stopped work on Tuesday evening pending a settlement of the dispute.[35] There was, however, no settlement in view and very little room for manoeuvre.

McLaughlin, as spokesman for Siemens, conducted his own public relations campaign, announcing that he had more applications for unskilled labouring jobs than he could handle and that men were being taken on at the rate of forty a day.[36] His colleague McGrath announced that the ships might be unloaded elsewhere, perhaps in Foynes.[37] Dockers in Foynes immediately held a meeting and indignantly rejected any suggestion of blacklegging and informed the Director of Labour that no boats for Siemens would be unloaded at Foynes.[38]

The following weekend saw the setting up of a strike committee in the city, composed of representatives of the National Executive of the Labour Party, Limerick Trades' Council and the ITGWU. The committee announced that it was ready to negotiate if the contractors were prepared to do so.[39] John Hickey was elected secretary of the committee and its headquarters were at the ITGWU office in O'Connell Street.[40]

Another mass meeting was held at the O'Connell monument on Sunday, 11 October. Three bands led the thousands who marched in support of the strikers.[41] Again the leaders of the Irish TUC and the Labour Party Executive were present and called for continued support in 'the biggest fight ever to face the labour movement in Ireland'.[42] One of the most interesting speeches of the day was made by Tom Irwin, secretary of the Operative Plasterers' Society and a member of Labour's National Executive. Referring to the German mechanics who were unloading the *Arabia* as 'scabs', he said that he had contacted the German trade unions and would have news for the mechanics shortly.[43] Within the week *The Voice of Labour* published a statement issued by the German Constructional Workers' Union, Hamburg, blacking the Limerick job and calling on all German workers on the Shannon Scheme to make common cause with their comrades on strike. 'Are you not ashamed to impose a sweated wage? Few rural districts in Germany would pay 66.5 pfennig for a 48-hour week. You will have to face the consequences in Germany when you come home!' This appeal to the German workers to support their striking Irish fellow workers was not successful. Given the language barriers, the geographical difficulties and the nature of the contract, it could hardly have been otherwise.

The following few days brought no new developments in the month-old dispute. Reports grew more pessimistic, as hopes for a speedy solution faded. *The Limerick Leader* described the prevailing mood: 'Impatience, bordering on disgust, seems to be the pervading feeling in Limerick in regard to the prolonged deadlock in connection with the Shannon Scheme ....'[44] There were various calls for public bodies, such as the Limerick Conciliation Board, to intervene. A statement made by the ex-servicemen unaffiliated to the Transport Union accused the ITGWU of dragging its feet in the whole affair and called on the general public to form an arbitration committee to bring about a permanent settlement. It was pointed out that men from all parts of Ireland were still being taken on as labourers and were working for 32s. a week. The ex-servicemen also stated that they were called off the site, even though they were being paid 50s. a week. The statement recalled the old Limerick prophecy: 'The stranger will flourish, the native will perish.'[45] The fact that the unaffiliated ex-servicemen were not receiving strike pay made them more anxious than most to return to work.

By the end of the week Limerick Corporation had yielded to the public demand for action. It was announced that Mayor Paul O'Brien and two other officials of the Corporation would form a committee to intervene in the dispute. Ironically enough, at the same time, the Corporation also announced that it was no longer going to supply water to the German ships.[46] News of the Corporation's committee was well received in the city. Even Casey's unaffiliated ex-servicemen, who were champing at the bit to return to work, said that they would defer their decision to return pending the Mayor's peace efforts.[47] But hopes of a settlement vanished when, because of the vested interests of some of the Corporation members, the Mayor's committee never really got off the ground.

Another arbitrator had begun to actively interest himself in the dispute. Father J. Cleary CSSR, spiritual director of the Redemptorist Holy Family Confraternity,[48] had already been in touch with the various parties in the dispute and was hopeful that a conference might be held in the near future.[49]

These moves gave rise to fresh prospects for an end to the strike. But by the weekend these hopes had been shattered. First of all, coal workers refused to deliver coal to Siemens.[50] Then, on 18 October, the Vice President of the Executive Council, Kevin O'Higgins, stated in Dungarvan that a secret society was fomenting trouble in the Shannon Scheme. He said that the organization's aim was 'to combat certain anti-British trade tendencies' which, he claimed, were showing themselves in the Free State. He alleged that an ex-army officer was the ringleader of this group, and warned those occupying pivotal positions in labour bodies or in organized associations of ex-servicemen, British or Irish, to watch out for agents of secret organizations. O'Higgins stated that the wage difficulty at the Shannon Scheme was directly due to a conspiracy by this secret society.[51] This would seem to be more a case of paranoia on O'Higgins' part, rather than a statement based on hard fact.

Needless to say, the Minister's Dungarvan speech did not help to soften the strikers' attitude towards the Government or Siemens. The final blow to the remaining slender hopes of a settlement was a letter from Siemens to Father Cleary. The contractors stated:

> We do not understand what right the representatives of Limerick city labour have to refuse to accept rates in Limerick city which they admit to be fair rates on the alleged grounds that other rates, which do not concern them, and which are higher than those rural workers receive, are too low. In the circumstances outlined, the firm does not see that any useful purpose would be served entering into negotiations with a negotiating committee which represents Limerick city workers with whom, as far as the firm is aware, there are no points at issue. It is regretted therefore that the firm cannot meet in Conference with the negotiating committee.[52]

The letter spelt out the worst fears of the strikers, but caused little surprise.[53]

McGilligan, the Minister for Industry and Commerce, was in the United States all this time. He had gone there to examine the control and structure of the country's electricity industry. His department, however, had remained remarkably

■ Travelling from the docks through Limerick city with an enormous turbine rotor.

silent on the dispute, even though it had come under severe criticism from several quarters for having advised the contractors to set the weekly rate for labourers at 32s. On 20 October the Department wrote to the National Executive of the Labour Party alleging that there had been an irregularity and that no formal complaint of a breach of the Fair Wages Clause had been lodged with the Department. The letter also pointed out that

> there would appear to be evidence that your Executive is endeavouring to establish some new principle to be applied as a test to the wages paid under a Government contract other than that hitherto followed, the general standard of wages ordinarily paid by good employers in the district where the work is carried out.[54]

Two days later, on 22 October, a meeting took place between government representatives and officials of the National Executive of the Labour Party. The government repeated its complaint of the irregular procedure and professed ignorance of any matter in dispute other than the weekly rate of 32s. which, it claimed, compared favourably with the wages of farm labourers, road workers and similar workers needed for the job. The government admitted giving information on wages to the contractors, and therefore accepted responsibility for the wage; consequently it was prepared to enter into discussion as to whether or not it was a fair wage. The government objected, however, to discussing the matter with the negotiating committee in Limerick where it thought the atmosphere would not be helpful.[55]

The Labour representatives proposed that negotiations on rates and conditions for unskilled labour and other questions, not purely local, be conducted nationally with the National Executive, and that negotiations for skilled workers in any particular locality be conducted with the local representatives of the unions concerned, or their headquarters, or with a local negotiating committee.[56]

The government voiced no objection to the proposal of the National Executive and said that it would submit the proposal to the contractors. On 26 October the National Executive of Labour approved the action taken by its representatives during their meeting with government.[57]

On the following day, 27 October, Siemens effectively torpedoed the proposal of the National Executive by stating that 'the firm had no objection to discussing with the Transport Union at any time matters which affected its members, but it reserved to itself the right to discuss with other Unions or Associations, for example, Associations of National ex-servicemen, etc., who may be prepared to offer their labour'.[58] On the same day the unaffiliated ex-servicemen returned to work.[59]

On 28 October the National Executive requested a further meeting with the government and suggested that the contractors be represented. The Department replied that since the ministers responsible were out of the country, no useful purpose could be served by further meetings.[60] Labour had no option but to suspend negotiations until McGilligan returned from the United States.[61]

A noteworthy feature of the dispute at this stage was the manner in which both parties conducted the campaign. McLaughlin, the Siemens' representative, had his almost daily report in the national press of 'steady progress being made' in spite of delays, and of new men being taken on. Labour conducted its own campaign, mainly through *The Voice of Labour*. Under the headline 'No Moleskin Joes', a correspondent from Ardnacrusha wrote that not one among the 150 navvies there could claim the slightest resemblance to one: 'They are the usual material of which scabs are made – down-and-outs, and jailbirds who never worked.'[62]

The newspaper also carried the words of the song 'Thirty-two bob!', which was doing the rounds of the pubs of Limerick and Ardnacrusha at the time:

Thirty-two bob! Thirty-two bob!
Come and we'll give you a beautiful job!
Come and enjoy some light recreation
Down on the Shannon electrification!
Sit down at once and send in your name,
And start in at playing an elegant game;
All you've to do is spend a few hours
Admiring the sun as it shines through the showers,
While you're up to your waist in mud and in stink,
Wielding a shovel or staking quick lime,
Shoving a barrow or lifting a load,
Digging a channel or making a road.
We don't want to strain you,
And so we won't detain you
For more than a mere fifty hours on the job,
And for that we'll pay you thirty-two bob!
Even that's not the might of our generous care,
We've taken great pains to ensure your share
Of good things to eat –
A whole half-pound of meat
And some scones at mid-day
And morning and night bread and jam with your tea:
For this we will charge you no more than twelve bob.
Sure everything's cheap on this wonderful job!
If you want any supper all you've to do
Is ask for it nicely and pay for it too.
After that you can roll yourself up in a cot,
In a nice wooden cabin set up on the spot.
You won't have to worry about paying rent,
For a rig and a stretch we won't charge you a cent.
Now isn't that a really beautiful job?
And don't forget a whole thirty-two bob!
Why don't you jump at the chance of your life?

Did I hear you say something about having a wife
And kiddies at home who will want to be fed?
What's wrong with you is, you've got a swelled head,
You shouldn't have children, your wife should be dead.
Didn't you know that the new Irish nation
Is only to last for one generation?
How dare you be married when your rate for the job
Is thirty-two bob! Thirty-two bob![63]

On 3 November the Dáil resumed, and Deputy Thomas Johnson, leader of the Labour Party, vehemently attacked the government's policy. He moved a motion that all steps should be taken to ensure that the men be paid a living wage before any more work was carried out on the Shannon Scheme. He argued that the wage was a sweated wage and that it did not allow for increases in the cost of living. He asked if they intended to build a new Ireland on unmarried men? The government rejected the allegations of Congo conditions and coolie labour and claimed that wages and conditions compared favourably with those of similar labourers. The Labour motion was defeated.[64]

The government's claim that it could not intervene and that the project would not be viable if the wages were raised was strongly supported by farmers who feared that higher wages would unsettle their own workers.[65]

The defeat of the Labour Party motion in the Dáil was a crucial blow for Irish labour in general, coming, as it did, only four years after the foundation of the state. And, as the debate was in progress in Leinster House, work was progressing on the site at Ardnacrusha, albeit slowly. It soon became clear that the government was not going to intervene. Labour therefore had to do something drastic, if the striking workers were to succeed.

Rumours had begun to circulate in Limerick of a general strike threat.[66] Then, on 9 November, the National Executive of the Labour Party summoned a special meeting of unions and the Limerick Trades' Council for 16 November in Dublin to discuss the deadlock. The special conference declared a rigorous boycott of the Shannon Scheme and called on all workers to treat the site as unclean and untouchable. Delegates blacklisted the men working there and appealed for financial aid for the strikers. Another important decision taken was the calling of a special meeting of the Trade Union Congress for Monday, 30 November.[67]

■ Passing safely by Limerick's Treaty Stone at Thomas Bridge en route to Ardnacrusha with turbine rotor.

McGilligan had just arrived from America and replied immediately to the union's boycott threat:

> The National Executive seems to have discovered in the proposal to provide work for some thousands of labourers on the Shannon Scheme, at better wages than they earn today, a veiled attack on the wage standard. That no standard can be durable except the standard of what the employment can afford they must surely have learned that this discovery of an entirely fictitious "attack" does more credit to their morbid suspicion than to their common sense or to their regard for the present or future interests not only of the nation but of the whole community of workers. Let the National Executive stand out of the way of those who desire to see the country develop according to the reason of its real resources and who are prepared to work hard to secure that object.[68]

As if the Minister's out-of-hand dismissal of the union's action was not bad enough for the morale of the strikers, the week of 15 November saw them taking another body blow. Members of the ex-servicemen's association affiliated to the ITGWU, led by ex-Brigadier Tim Murphy, decided to return to work, in spite of their union's official position.[69] The dockers also met, but decided to continue with the strike. However, it was reported that many of them were in favour of returning to work.[70] Michael McCarthy, the local Transport Union organizer, immediately announced that, despite the apparent change in the attitude of some of the workers, there was no change in the union's policy. The position had altered slowly but significantly, however, and it was reported in the Dáil that over 600 men were then working on the Shannon Scheme.[71]

Sunday, 22 October, brought another rally at the O'Connell monument in the city. Five thousand people attended to hear Thomas Johnson, Labour leader, and other Dáil deputies address them.[72] Cork also had a mass meeting in support of the strikers, and there were reports too of rallies in other parts of the country.[73] Transport Union branches throughout the country, from Waterford to Sligo to Dublin, now began to contribute regularly to the Limerick strike committee fund.[74]

On Monday, 30 November, 130 delegates from the Free State and Northern Ireland attended a special meeting of the Trade Union Congress in the Mansion House, Dublin, which was heralded as the most important special conference since 1918. The meeting endorsed the boycott of the Shannon Scheme, and called on the government and contractors to reconsider and negotiate. Delegates condemned the 32s. wage and resolved to protect the trade union rights which were threatened by the dispute.[75] The decision of Congress to boycott the scheme could have affected 250,000 workers in the country but, in the event, it was never implemented.

The Senate debate took the same bitter turn as the Dáil debate. Séamus O'Farrell, moving the motion on behalf of Labour, urged the government to recognize 'the right of workers to rates of wages at least sufficient to provide them and their families with the indispensable necessities of civilised life'.[76] He stated that Labour was in favour of the Shannon Scheme, but that the conditions pertaining and the wages offered gave him no option. He added: 'Instead of creating a Gaelic Ireland these conditions will create an Irish China.'[77]

The Earl of Mayo painted a picture of 'navvies frying their beef-steaks on a shovel,' and could not see the reason for all the fuss.[78] He appealed to Labour to take a broader view: 'I hope the Labour Party will not really interfere in this serious matter, because it is a serious matter to encourage men who are only too anxious to make a little mischief.'[79] The Labour motion was defeated.

At this stage it was becoming increasingly evident that the unions had lost the fight. At the local level, however, the boycott was maintained and intensified. Dance-halls, picture houses and the Markets Field were closed to scabs. The first

■ To Ardnacrusha via Longpavement, with generator body in tow.

list of scabs was sent to all labour halls in the city. Top of the list were the names of the two ex-servicemen – Tim Murphy and Terry Casey.[80] A second list followed a week later.[81]

The dispute dragged on over the Christmas with very little of note occurring, except some occasional attacks on the Germans.[82] Meanwhile, other issues, such as the 'Boundary Crisis', had begun to occupy the public mind. The press interest waned and little or no space was at this stage being given to what had become known in the newspapers as 'The Shannon Scheme Crux'.

On Monday, 11 January, pickets did not appear at Strand Barracks, Longpavement, Ardnacrusha and the Siemens' office in Limerick. A prominent Labour leader announced that the strike was as dead as Julius Caesar, but the strike committee said that the boycott was still in place.[83] On the following Friday, 15 January, the dockers, who had played such a vital role in the whole dispute, decided to return to work on the old conditions.

Building work on the Shannon Scheme now began in earnest, and by the middle of February the first unit of electricity was produced in the temporary Power Station at Ardnacrusha. There was no official announcement from the Transport Union that the strike was over. The only indication of how things stood came in a letter to *The Limerick Leader* on 2 February, from M. O'Shea of the strike committee, calling on everyone to build a 'greater Limerick':

> If Limerick is all wrong
> Who's to blame?
> If the people all starve on
> Who's the shame?
> Will you help this wrong to right?
> Come along and spread the light
> And do your share.

After this forlorn plea, the strike gradually petered out, but much bitterness remained for years afterwards.

Though the strike has been neglected in labour history studies, a reference to it was made in 1979 in the Electricity Supply Board's publication to mark the fiftieth anniversary of the completion of the project.[84] The anonymous writer gives a brief outline of the cause of the dispute from the Board's point of view and concludes with an explanation for its failure:

> Lorries bringing material to Ardnacrusha and Clonlara each carried an armed soldier. Ships were unloaded at the Docks under the protection of soldiers with fixed bayonets. This incensed the dockers, who occasionally used to pelt stones at workers bringing material up the Shannon by boat, or try to pull them out of the boats with grappling hooks. On one well-remembered occasion, when the army escort was absent for some reason, the dockers made a move to throw a light engine, which had just been unloaded, into the river. The foreman in charge of the ex-servicemen who were unloading the ship, an ex-army captain, immediately drew a revolver, and pointing it at the dockers said: 'The first person to touch that engine is a dead man.' On this occasion the dockers had no option but to withdraw quietly. This tough approach by the ex-servicemen, and the heavy police and army protection, ensured that the strike had no chance of succeeding in Limerick city. It had even less chance of success in Ardnacrusha and the surrounding district. The

country was going through a severe economic depression and work was scarce. People were only too anxious to work for the wages being offered. Many workers lived in the area. Others came from as far away as Scotland, and there was a large contingent from the West of Ireland. In all these circumstances the strike was doomed to failure from the start. An additional factor was the feeling that Siemens would have been willing to pay more, but were forbidden to do so by the Government. Early in 1926 the strike fizzled out, and no more labour troubles occurred for the duration of the Scheme. As one newspaper editor put it at the time: The strike was by men who themselves had no complaint to make on behalf of men who made no complaint.

The Shannon Scheme strike has long been forgotten, but the struggle by Limerick workers against low wages is worthy of a place in the annals of Irish labour history.

3 | The Land Question

When one examines the Census returns for south-east Clare for the early part of the last century, farmers frequently registered themselves as 'Republican Farmer', under profession. In the townlands of Ballykeelaun, Ballyglass, Castlebank and Lakyle where the Power Station was built, many farmers were 'Republican Farmers'. The term underlined that powerful link that existed between nationalist hopes and a farmer's passionate attachment to his land. It also resonated the dynamic for agitation which that link provided in times past.

In post-independent Ireland these same people just registered themselves as 'Farmer'. This is not to say that the bond between a man and his acreage was less than in colonial times. Far from it. The numerous disputes about title and rights of way resulting from Land Acts proved that. But one effect of the 1923 Land Act was to make the smallholder more secure and settled. He also saw the Cumann na nGaedheal government largely ignore the pleas of the 'landless men' and 'relatives assisting', thus eliminating any possible further disruption or division. He could now relax in his title and comfortably wear whatever pressure the government put on for more efficient production. For the farmer it was now a case of 'what I have I hold' – no change, echoing in a way (even though he would not like to admit it), the motto of that renowned 'Gentleman Farmer', John Singleton of Quin, whose family crest bore the words *Mutare Sperno* – 'I Spurn Change'.

■ Excavators at work in Ballykeelaun digging foundations for the Power House.

Years later, after independence, in the locality of Parteen, Ardnacrusha and Clonlara, people still sang the eviction song of an earlier generation:

> Wait 'til the landlords go, Paddy,
> Wait 'til the landlords go,
> Hold fast the sod and trust in God,
> And wait 'til the landlords go.[1]

The landlords had certainly gone and dispossession was just a bitter memory, but now there was a new challenge coming down the tracks. The native government was proposing to move into south-east Clare and dispossess entirely some farmers of their land, to make other farms unworkable, and to disrupt profoundly the economy of all holdings affected by the Shannon Scheme.

The Shannon Electricity Act of 1925 empowered the government to take whatever land and property it needed to execute the project. Surprisingly, there was little or no dissent in the Dáil and Senate to its passing. And while farmers were to experience massive change because of it, neither was there any major dissent from them. They too bought into the project and ceded their lands without protest or coercion. This was an extraordinary development and a remarkable achievement for the government in terms of marketing the initiative and in giving people ownership of it at an early stage.

It is difficult to know to what extent detail of the debate on the Shannon Electricity Act in Leinster House percolated through to local landowners. The Minister for Industry and Commerce, Paddy McGilligan, was uncompromising in his attitude and very tough. For example, Section 4, Subsection 2, proposed to give the Minister authority to enter and take possession of any lands or premises, subject to giving one month's notice in writing to the occupier. In the Senate, Sir John Keane argued that this was too harsh because of the hardship it would cause to individuals and their families if they were turned out of their homes with nowhere to go. It was somewhat ironic that it was Sir John, one of the biggest landlords in Munster and a Unionist to boot, who gave the young Free State Minister a lesson in compassion towards the smallholder. He wanted a three-month notice to be the norm. The Minister refused, saying that this would cause too much delay for the contractors. He declared that maps and plans would be published and that people would be informed as to their future. He stated that compensation was there for the people to claim and that the government had no intention of building houses to replace those taken over or knocked down! Fortunately for occupiers Keane's amendment was accepted.[2]

The occupation and valuation survey of land to be acquired for the Scheme was a very complex and onerous operation. It involved the Land Commission, which, as a rule, could provide particulars of ownership and indicate the boundaries of the holdings affected. In cases where the Land Commission could not do this, the information had to be got by inquiries on the ground. Maps were drawn up indicating the 'limits of deviation' of the area required for the project. Within those 'limits' the Land Commission marked the boundaries of the various holdings needed. These maps were then given to a team of inspectors for verification. When they had completed their job, the maps were passed on to the Ministry. Each map was accompanied by a 'Book of Reference' that contained the names of owners, lessees or occupiers, rights of way and anything else considered relevant. Each holding was numbered.

In a way, the inspectors were pivotal in the process of land acquisition and

■ Excavator discharging rock into tipper wagons for use in headrace banks.

they were kept informed by the Ministry as to which area was prioritized for immediate excavation, blasting or whatever. If the Ministry needed information urgently on any section of land, the inspectors were the ones to provide it. It was they also who provided a confidential memorandum to the Minister in October 1925, outlining for him various points in the Bill relating to the acquisition of land and advising him on how to proceed where difficulties arose.[3] It was quite a lengthy document and very detailed in its advice. For example, they counselled that it was necessary to value not only the lands within the 'limits', but also adjoining and nearby lands that might be affected. In some cases, an entire holding had to be valued in order to value accurately the part to be acquired because the acquisition would impact on the annuity paid to the Land Commission.

In order to ensure uniformity in valuation, the inspectors worked from a table containing basic prices per acre for different types of land – meadow, arable, bog, scrub etc. They also had to take into account where the land was situated; for example, if the land was close to a town, adjustments would have to be made in costings. When the inspectors were making their valuation they had to note the system of farming followed on each holding – the carrying power of the land, the area under tillage, etc. The owner of the land had to make a general statement regarding the loss or injury he was likely to suffer as a result of the property being acquired.

The main factors which really determined compensation were if the property was being temporarily acquired (for part of or for the whole duration of the Shannon Scheme); if there was permanent acquisition; or if there was to be an exchange of lands.

The first option of temporary acquisition was simply a matter of fixing an equitable annual rent for the period during which the premises or land were occupied by the government. The only complicating factors here were if there were crops on the land that the farmer could not remove and, possibly the question of surface damage as a result of the occupation.

The second option of permanent acquisition was expected to be pretty straightforward. Even when the lands affected were a relatively small part of a more extensive holding, compensation would be on a cash basis as provided for in the Act.

The exchange of lands option was a more complex issue. In some cases, lands to be acquired could be portions of a small holding, of a holding already uneconomic, or of a building on which the owner was entirely dependent for a livelihood, or even portions of non-residential holdings. The question was, what should happen in those cases? Should compensation take the form of other lands? Should the government acquire the entire holding and use the balance not required by the contractors for the purpose of giving exchanges to other smallholders?

It was felt by government officials that this second option could be fraught legally and that it would not solve the problem of compensation entirely, as there would still be cases in which such lands were not available. In these cases where the tenant was losing his/her dwelling house and offices, as well as the bulk of their land, it was proposed that the only satisfactory solution would be migration. The question then arose as to whether the Land Commission could provide untenanted lands elsewhere within a reasonable distance for those dispossessed.

Several such smallholders had already expressed the wish to be compensated by obtaining a holding elsewhere. The government also favoured this system of compensation as it could save it a considerable amount of money. But apart from the cost saving, it appeared a more equitable way of compensating those so affected.

A particular difficulty arose for the government in areas where the contractors could not say for certain what land was to be acquired for temporary occupation and what was to be acquired permanently. To get around this, it was proposed that, in cases where a relatively small parcel of land in an extensive holding was affected by the works the farmer would be paid a rent until such time as it was decided to take the land over permanently or temporarily. The thinking behind this proposal was that the farmer's operations would not be affected to any great extent by the occupation and that the rent would compensate sufficiently for any minor losses incurred. Needless to say, if the decision was to acquire such small areas temporarily, then there was question of compensation in respect of surface or other damage before the premises were given back to their owner. In the event of permanent acquisition compensation was assessed on a cash basis.

There were other cases where the whole or a major portion of a smallholding was involved and the question of temporary or permanent occupation hung over them. In these cases the owners were in an invidious position, caught between a rock and a hard place. The contractor was occupying the farm or the premises and the owner was getting no return from the property. To avoid such hardship it was suggested that the owner should be offered the choice of either receiving compensation immediately – as for permanent occupation – or, alternatively, of receiving an annual sum by way of rent out of the lands, until such time as it may be possible to decide what portion of the lands, if any, would be restored.

The final point in the inspectors' memorandum dealt with those dwelling houses earmarked for demolition. In some cases, where their homes were to be knocked down, the occupants of those houses were not always the people who would receive compensation under the Act. Consequently, they would be placed in a position of considerable hardship if some provision were not made for alternative accommodation. The inspectors pointed out that their case merited attention and suggested that a reinstatement of such buildings might be made a condition of the compensation awarded.

On 2 September 1925 the nineteen Books of Reference containing the maps and plans indicating the lands and properties to be acquired by the government were deposited in Limerick and Dublin for inspection by the public (Appendix 1). *The Limerick Leader* reported that just 43 acres (amounting to three holdings) on Castlebank, Ballykeelaun and Lakyle (in the electoral area of Ballyglass) would be affected by the works immediately.[4]

The Books of Reference were drawn up by a Reference Committee that also appointed arbitrators who would adjudicate in cases where agreement could not be reached with property owners. They could hear up to seven or eight cases a day, when issues would not be too complex. There were quite a few arbitrators throughout the Scheme and after it, particularly dealing with cases that had become intractable and messy. The best known of the arbitrators were Messrs O'Malley, Murphy, Martin, Hewson and Sharp Bolster.[5]

The public face of the arbitration process was Arthur Taylor from Clontarf in Dublin. He was appointed by the government as a valuer and he travelled to the various areas affected by the construction works to negotiate with farmers and property owners. He became well known throughout the length of the Scheme and apparently enjoyed the trust of the various property owners he met. He was a tremendously skilled negotiator, as is evidenced from the number of agreements reached by him. However, he had his difficulties, as he noted when he was trying to get accurate information from farmers whose lands at Birdhill had been affected by the Kilmastulla River that had overflowed. In his report to the Secretary for Industry and Commerce, he noted: 'One day on going on to the land a horseman rode across the meadows and instructed several parties who were cutting their meadows to give no information to me.'[6] He was permanently on the road visiting, and sometimes re-visiting, people scores of times, attempting to reach a settlement with them. His challenges ranged from farmers to fishery owners, from cottagers to absentee landlords. Everybody wanted top price and it was up to him to deal with owners' frequently inflated notions of what their property was worth. He once commented that the people around the area would 'swear a hole in a pot' in order to get a better deal. But having reached a settlement, his biggest difficulty then arose – getting the government, and specifically the Department of Finance, to pay the money.

It became apparent very early on in the Scheme that the tardy processing of payments by Finance, or in many cases failing completely to do so, was a major problem, adversely colouring the whole project. Gordon Campbell, the Secretary to Industry and Commerce, identified this difficulty from the outset and complained bitterly to his Minister, Paddy McGilligan. He also gave Joe Brennan, the Secretary to Finance, a bit of his mind and left the latter in no doubt as to procedure and protocol. Writing to Brennan in June 1926 he baldly pointed out: 'Serious complaints are being received as to the delay in proceeding to settlements which may have a prejudicial effect on the whole Scheme.' Just in case Brennan

■ The weir at O'Brien's Bridge mid-way through construction.

did not know the procedure that had been adopted by Industry and Commerce in acquiring property, he continued and spelled it out very clearly for him.

> Particulars of the property acquired are furnished to the valuer, Mr Taylor and he to the Chief State Solicitor. Mr Taylor, after a visit to the locality, reports as to the measure of the compensation he recommends should be paid for the rights which the Department has acquired, on the assumption that the persons who own such rights are the persons who represent themselves to do so. Mr Taylor's reports are forwarded to the Department for concurrence in the payment of compensation according to that measure.

He then went on to outline the role of the Chief State Solicitor in the process as regards enquiries into all questions of any title.

> The Chief State Solicitor explains that he cannot get any assistance from local solicitors in furnishing title until he is in a position to indicate that he is prepared to pay so much money for the interests acquired …. I am to press for the concurrence of the Minister for Finance in the measure of compensation recommended by Mr Taylor … so that the Chief State Solicitor may be enabled to put forward subsequently a complete statement showing in detail the basis on which each case is to be finally disposed of.[7]

A very sheepish and compliant reply was received from Joe Brennan within two days: 'In reply to your minute of the 24th instant relative to compensation for the acquisition of property for the purposes of the Shannon Scheme, I am directed by the Minister for Finance that he agrees to the procedure outlined therein.'[8]

Perhaps Brennan and Finance's foot-dragging may have been due to the severe snub they received early on, when Paddy McGilligan announced at a Cumann na nGaedheal meeting in Mayo that the government was going ahead with the Shannon Scheme without informing Brennan or his officials in Finance beforehand. Brennan, the most senior of civil servants, was shocked and angered, understandably so. This led to an inter-departmental row and bad blood between the two departmental secretaries, as is evidenced in their correspondence and in Brennan's complaints to Blythe, the Minister for Finance, about Campbell's meddling in financial matters that were no concern of his. But Campbell had claimed and proved that Finance was not just obstructive towards the whole project, but ran the risk of damaging Ireland's credibility by not proceeding with the bond for the electrical contract without which work could not proceed.

The sharpness of Campbell's correspondence to Finance was prompted by the holding up of three cases of compensation in the Ardnacrusha area – that of Robert Holmes of Castlebank, Thomas Hartigan of Ballykeelaun, and water supply for Patrick Boyle and John Hamilton of Ardnacrusha.

In the case of Robert Holmes, his entire holding at Castlebank was 112 acres, of which 14 acres had been acquired permanently and 40 acres temporarily. On the 40 acres, the contractors had erected their temporary power station, residence for staff and workshops. The Power Station as constructed was a solid affair and, in the opinion of Dr Sothman, the Chief Electrical Engineer, it would be invaluable as an adjunct to the Scheme if it could be bought cheaply when construction was finished. The making of the foundations involved heavy excavation and the land would never be restored to the owner in its original condition. Furthermore, operating staff in the future could use the Villa residence. Mr Taylor and the engineers recommended that the forty acres be taken permanently. The Minister approved.

In the case of Thomas Hartigan, 80 per cent of his holding was acquired permanently and 20 per cent temporarily. His was a non-residential holding and as he was a large farmer the acquisition of the extra 20 per cent would not seriously affect his income. Taylor recommended that this would be cheaper than providing a water supply for the land. Again, the Minister approved.

In the cases of Patrick Boyle and John Hamilton, both parties would be deprived of water for domestic use and for livestock. The valuer recommended that it would be cheaper to lay a pipe providing water rather than allowing compensation for loss of water. This opinion was also acceptable to the Minister and he approved it.[9]

However, in spite of the Minister's approval and Gordon Campbell's pressure on Finance, payments were still held up. For some inexplicable reason the plethora of bureaucracy and red tape in the Department slowed down the money coming through to individuals thereby increasing frustration and hardship.

But the work had to go on. In addition to land to be acquired permanently and on a temporary basis, there were also cases of extra land being taken over during the construction because of special problems. For example, at Birdhill fourteen acres were taken for large embankments to prevent low-lying areas from being flooded.[10] There were the numerous rights of way to be negotiated and agreements to be reached with landowners for the erection of poles to carry telephone lines and electric cables. On average they got 5s. per annum compensation per pole so that some did quite well out of it. There were those, such as Patrick McCormack in Castlebank, who opted for a supply of electricity as compensation for the damage done in erecting and maintaining poles on their land. And there were those, such as Michael O'Reilly of Ballyglass, who allowed the poles to be erected free.[11]

In Clare alone there were nearly a hundred occupiers from whom land was acquired. The total acreage of the Scheme was 14,515.[12] By the beginning of 1928 fifty of those cases had agreed compensation but payment was withheld until occupiers could furnish proof of title to the property. In fifteen cases payments of compensation already made amounted to £12,012. That made an average payment of £800 in those cases. At the time that was good money.[13]

The first case of possession under the Shannon Electricity Act happened in May 1927, when an application was made in the High Court for the lands of Michael McNamara of O'Brien's Bridge. But this was the exception.[14]

In the remaining cases for compensation, the Minister for Industry and Commerce, in a written reply to a query from Eamon de Valera, said: 'settlements will probably be reached in all cases except perhaps a small percentage where the occupiers present demands are unreasonable'.[15]

Locals, however, thought that it was the Minister who was being unreasonable. In September 1926 the Clonlara Farmers' Union, under the Chairmanship of

■ The giant bucket dredger in action in the headrace, loading some of the thousands of wagons used in the project.

Arthur Stritch, severely criticized the government for not paying compensation for land taken some twelve months earlier, particularly as the County Council was still demanding rates be paid![16] Councillor Pat McMahon of Blackwater took up this latter point at the County Council meeting, where he pleaded for some latitude for his constituents. The reply of the Council Chairman was an emphatic 'No'. That did not solve the problem though; in August 1928 the Clare Board of Health received a letter from Mrs Nora Keegan of Ballykeelaun stating that all her land had been taken, and having received no compensation, she was still being compelled to pay rent for what she had not got. The Board decided to waive the rent as her situation was intolerable.[18]

It was not just the County Council that was vexed by the problem. So were the Courts. John Hastings of Ardnataggle (and a number of others) was sued for rates of £28 18s 8d at Killaloe Court in August 1928 by Denis Vaughan, the rate collector. Hastings' lands had been acquired with no compensation paid. A decision for the full amount was granted against him. The poor man was up in court again in May of the following year and the full amount was granted on that occasion also. It was a case of live horse and you'll get grass![19]

A neighbour of Hastings, John Daly, was summoned to Limerick Civil Bill Court where the Land Commissioner sued him for 15s. 4d. in respect of annuities.

Justice Flood adjourned the case for three months, declaring that the country had gone crazy – one department of the government taking land without paying compensation while another department sued for annuities.[20]

By mid-1929 some semblance of common sense had crept into matters. Peter Mulqueen of Clonlara was dragged into Limerick Civil Bill Court by the Land Commissioner, who sued him for one land annuity instalment. Like all the others, his land had been acquired for the Shannon Scheme but there had been no compensation. The case was 'Nilled'.[21]

At the same time, the Minister came under pressure in the Dáil from Paddy Hogan of Labour as to why the Arbitration Court, as provided for by the Shannon Electricity Act, was not set up. McGilligan replied that he was proceeding by agreement with the landowners. He said that notice of acquisition was served on 96 occupiers; agreement had been reached in 60 of those cases at a cost of £37,200 and a further £767 per annum was being paid for land acquired temporarily. The Minister added that in the 36 remaining cases there was no agreement as yet, but that 6 were on the point of settlement. He pointed out to Hogan that settlement by agreement had proved satisfactory up to this and that a case for arbitrating a claim had not been made so far.[22]

That was the view from Leinster House. The view from Browne's pub in Ardnacrusha was different, as the local savants had it:

> On the green banks of Shannon
> Since Siemens came nigh,
> No poor Irish farmer is happy – and why?
> The government to them compensation won't pay
> They've even lost faith in their friend J.J.[23]

J.J. was J.J. McElligott, Secretary in the Department of Finance and successor to Joe Brennan.

In 1932 a number of compensation cases still remained to be resolved. McGilligan listed these as 'difficult cases' where landowners, for one reason or another, would not settle. So at the beginning of the year, with an election in the offing, he made a determined effort to expedite many of these and succeeded in doing so.

The most poignant case was that of Patrick McCormack of Castlebank who had died a few weeks before. According to Arthur Taylor's report attached to the

■ The new wooden village of Ardnacrusha with Irish and German camps and the foundations of the Power Station in the foreground.

agenda for the meeting of the Shannon Board of Control that considered compensation payments, McCormack had suffered a lot during the works and was soured as a result. His land was just at the site of the Power Station, where he had fifty milch cows and other livestock. His house had been badly damaged by the blasting and two wells supplying water for domestic purposes and for the animals had dried up. As a result he had to travel one-and-a-half miles for water for domestic purposes for six years. Moreover, a stream that flowed through his land had been badly polluted by sanitary arrangements at Ardnacrusha camp. The whole situation was practically unbearable for him and his family. For example, on one day alone during blasting, seven of his cows had 'slung' their calves. McCormack claimed that the constant worry about further losses through the explosions and general disruption caused his health to break down seriously in 1928. Taylor had met him many times for discussion of his claim for £2087 (plus interest from 1925), but there was no agreement reached. The contractors had paid him £300 plus costs for damage done by blasting and flying rocks. But for the last three years of his life the poor man was an invalid. Now Taylor met with Mrs McCormack and her advisors but was very anxious that the case should not go to arbitration. He felt that had it done so the family had a very good case for heavy damages. Fortunately for the government, Mrs McCormack accepted £1,080 in full compensation, with no interest. Breaking the total down, this meant she got £400 for land permanently taken, £155 for disturbance and loss of amenities to their residence, £300 for loss of water and £225 for rent of twelve acres acquired temporarily. Needless to say the Minister approved the settlement.

McGilligan also managed to settle a number of other smaller claims on that occasion, the more significant of which were that of Mrs Marie Ryan of O'Brien's Bridge for £350 and twelve guineas in fees, and of Arthur Stritch of the pub in Clonlara for £340, plus four guineas in fees.[24]

And so, the flag changed. Fianna Fáil swept to power ousting the Cosgrave-led administration. Hopes were high among the tenant farmers of better settlements from de Valera. There were still a number outstanding and it was decided to have these heard formally in front of a government-appointed arbitrator.

But the mandarins of Finance still held sway and there was little change in the pattern of awards or even of administration. Arthur Taylor, who had been so effective in achieving agreements with landowners and, obviously enjoyed their trust, fell seriously ill and so was excluded from giving evidence at the formal arbitration process. It was decided by the Acting-Minister to replace him with a local valuer who would have knowledge of sales and the value of land in Clare. The man chosen was Michael MacMahon of MacMahon Auctioneers and Valuers, Ennis. He indicated his willingness to give evidence and to make the necessary valuations for an inclusive fee of five guineas in each case. 'The acting-Minister is not satisfied, from the information at present before him, that these terms are reasonable, and he accordingly proposes that an endeavour should be made to secure the services of Mr MacMahon at a fee not exceeding three guineas a day, plus necessary travelling expenses and subsistence allowance. It is proposed, in the first instance, to endeavour to secure the services of Mr MacMahon at a fee of two guineas per day, but the Acting-Minister would be glad to have the approval of the Minister for Finance for a fee not exceeding three guineas should it prove impossible to obtain Mr MacMahon's services at a lower figure.' As it so happened, MacMahon turned down the Department's offer and S.F. Ebrill of the Limerick auctioneering firm, who was happy enough with three guineas, took the appointment.[25] This was a classic manoeuvre by government officials at the time, to cut costs at every possible corner; it was applied to the hiring of people by the lower stretch of officialdom to the top civil servants.

It also applied, in the most curmudgeonish of terms, to claims of compensation. It was no wonder that some claims lasted long after the Shannon Scheme was opened in 1929.

While the Books of Reference and accompanying maps are easily accessible in the National Archives, the files relating to many occupiers are not available on recall through their file numbers at time of writing. For example, File SS 32 relating to the Hamilton family farm in Ballykeelaun or File SS 53 detailing transactions of the Dundas family, both of whom were dispossessed by the site of the Power Station, would make for very revealing reading. But, unfortunately, they are not there. And so it is with the files of so many families along the length of both canals. To date, an enormous amount of work has been done on the Shannon Scheme records, but perhaps with further refinement the wayward material will turn up, filling those missing gaps and providing the answers to so many questions.

4 | As Others See Us

Of the thousands of files in the National Archives dealing with the Shannon Scheme, there is one that gives a great insight into the relationship between the Irish and the Germans on the project. File SS13929 deals with Irish workers. It covers a span from January 1926 to December 1929 and deals with industrial flashpoints, various human-resource problems and allegations of racism against German foremen. It also covers the winding-up of the Scheme and political problems that arose as thousands of men were being thrown back onto the labour market.

In the aftermath of the bitter and protracted strike at the beginning of the Scheme, Irish management was particularly sensitive to the issue of wages and closely monitored the contractors. The German workmen were paid more than their Irish counterparts and, in fact, contractor management was not averse to paying more than Paddy McGilligan and the government would allow. This was a well-known fact throughout the length of the Scheme, but there was no way that the Germans were allowed to breach the government guideline. The gimlet eye of the Resident Engineer in the Strand Barracks was on all sites and he quickly took the contractors to task if there was the slightest breach of contracts or agreements.

Siemens Bauunion had their transfer station at Longpavement. Here they set up a railhead and used the base for stores. It was from here that over three dozen

■ Rail elevator lift for heavy turbine equipment.

trucks would trundle daily to the various sites between Ardnacrusha and O'Brien's Bridge, ferrying equipment and spares as needed.

On 27 January 1926 a letter arrived on the desk of the Chief Engineer, Shannon Power Development, Professor F.S. Rishworth. It was from Herr Heintze of Siemens Bauunion. In good Irish Civil Service style, it began with 'A chara', as did all their correspondence, closing with 'Sinne le meas'. 'I have been informed that you have taken displeasure at our running a stone-crushing plant at Longpavement and that you have therefrom inferred a breach of my promise with regard to the works at Longpavement.'[1]

Apparently the contractors had levelled the ground at Longpavement, thereby causing a small quarry. Subsequently they brought in a stone crusher and then used the end product for ballasting the light railway track from Longpavement to Quinpool Bridge. They had used their own electric current to run the stone crusher and a compressor. This, they argued, was much cheaper than getting supplies from Ardnacrusha, as the rock from there was just as urgently required in that locality.

All of this would appear to be eminently reasonable. The important point, however, was that the stone-crushing operation was absorbing labour at city rates. Seventy men were employed at 1s. per hour, which pushed up costs. Labourers at Ardnacrusha were employed at 8½d. per hour, so that stone production there would be much cheaper.

The government was insistent that the contractors employ as few men as possible within the city boundaries, because of costs. It ordered the immediate closure of the stone crushing plant at Longpavement and made it very clear to the Germans that such breaches of agreements would not be tolerated. A very contrite letter from the contractors concluded:

> By your wish we have immediately stopped the stone works at Longpavement although you have no extra costs on their account … I am very sorry that our measures which have solely served to improve the success of the works have made an unfavourable impression on you. I wish to apologise and hope to have now satisfactorily explained this matter.[2]

But these were early days and, according to the Germans, there were conflicting messages coming from the Irish side. Tommy Mac Laughlin and Joe McGrath of Siemens had met Rishworth and Gordon Campbell of Shannon Power Development in September, as a result of which the contractors decided on their numbers for the city, unaware that there was a limit on how many labourers could be employed there. It subsequently transpired that there was a stringent quota being applied by the Shannon Power Development team and even though this was strongly objected to and hotly contested by the contractors, they just had to come into line, much to their chagrin.[3]

One would have thought that by 1927 most major problems would have been sorted out and that efficient communication systems would have been pretty much in place. Not so. In January a group of Donegal men refused to work in protest at what they perceived as being hired under false pretences.

After the prolonged strike at the beginning of the Scheme, Joe McGrath set up a system of recruitment from labour exchanges throughout the country. One day men would arrive from Wexford and on the next from Mayo, Waterford or wherever.

In mid-January 1927, 44 men arrived from Dungloe, having been recruited through the exchange there. These were due to start work in Blackwater, but when they were presented with the 'Engagements to Repay Fares' form they refused to sign it. Eventually they did sign when they were told that they would not be able to start work without signing. Then they were brought to the camp at Ardnacrusha where they were accommodated. The next day they were brought

 J.K. Prendergast, resident engineer with Shannon Power Development.

to Clonlara where Dr McSwiney examined them. Afterwards, instead of reporting to the Pay Office in Clonlara, they went back to Ardnacrusha and informed Stapleton, the Camp Commandant, that they would not begin work as the conditions under which they were now being asked to work were quite different from what they had been told by the manager of the Employment Exchange in Dungloe. He had told them they would receive 35s. 5d. for a 50-hour week and a rate of 1s. 5½d. for an hour overtime, also that the railway fare would be paid by the firm, but this would be deducted from their wages at a rate of 2s. per week.[4]

The stand-off was quite an embarrassment for the office of the Labour Advisor, as McGrath's department was called, particularly as the memory of the strike still lingered on. An official was quickly dispatched to Ardnacrusha to interview the men and to sort out the matter. That he did, disabusing the men of their high expectations, and informing them of the going rate of 32s. per week for labouring work. At the time all of these men would have suffered from what was known colloquially as 'the want of money'. Most of them would never have had a weekly wage. Many of them would have worked as agricultural labourers and endured the hardship involved, as well as the vagaries of uncertainty of payment. The prospect of a few years of accommodation, a canteen where one could indulge in the luxury of eating meat daily if one wished, and money into the hand every week, was enough to placate the most militant of the Donegal men and to dissipate their protest. No more was heard of it.

J.K. Prendergast, the Resident Engineer in Limerick, was a man who had a broad portfolio covering all the engineering and logistical areas, as well as personnel issues. In mid-1926 he wrote a revealing confidential memo to his boss, Professor Rishworth, as to how things were going on-site:

> I understand there is considerable local gossip and discontent re the recent dismissals of Irish labourers and skilled men by the contractors. The local view as expressed to me by some influential men in the town, who are in a position to ascertain the facts, is, that if dismissals are necessary, the contractors should endeavour to avoid dismissals which cause undue hardship.[5]

■ All scaffolding was of timber, much of it fabricated on site in the workshops.

He went on to report having heard of labourers being dismissed who had stood by the contractors under great difficulties at the start of the Scheme and who worked right through the worst of winter weather; he had heard also of other men being retained who would not suffer any hardship even if dismissed.

Another issue that concerned Prendergast was that the contractors in many cases gave quite inadequate and sometimes unjust grounds for dismissal. For instance, he had heard of a crane driver being hired and subsequently being dismissed on the grounds of incompetence, although he was never allowed even handle the crane.

Prendergast maintained to Rishworth that much of this information had come to him through one of the contractors' employees, so that for the moment he would not like to carry it further. He declared that all he was doing was flagging Rishworth about rumours, discontent and ill-feeling at local level.[6]

Within a short time he had an opportunity to act, and he took it. He wrote to the contractors:

> I am informed that it is your practice not to allow Irish foremen to work unless more than five of their gang report, whereas there is no such rule for the German foremen. An instance has been reported to me where an Irish foreman and five of his gang reported for work after a rainy morning, but they were not allowed to

work and at the same time a German foreman in the same place and on similar work was allowed although only one of his gang was present.[7]

He told the contractors that he considered this to be unfair discrimination, that it would disaffect many good Irish labourers and foremen, and that he wanted the matter investigated.

The Germans replied on the following day dismissing Prendergast's charges: '...There is neither a rule nor a practice that Irish foremen are not allowed to work unless five of their gang report.'[8] On the day they reported that the German foreman was ordered to carry out urgent work and as there was no usual excavation being done the Irish foreman could not do his work. 'We must add that in a general way that we are anxious to employ the Germans wherever this is possible, for the simple reason that according to the Contract we have to pay their wages weekly and not daily as is the case with the Irish workman.'

The Resident Engineer was livid: 'I find that my facts are substantially correct and I want an explanation of those facts – not a denial of same.'[9]

The exchanges were sidelined by an incident at the weir in which an Irish worker assaulted a German engineer, quickly followed by the questionable dismissal of an Irish train driver who had been abused by his German foreman and called an 'Irish swine'. The driver had worked with the contractors from the start, even during the strike; he now feared that it would be impossible for him to work anywhere else because of his anti-union action then.[10] His dismissal led to heated exchanges with the contractors, they claiming that not only was the driver incompetent but so were 60 per cent of all Irish drivers. Naturally enough, the German foremen closed ranks with their colleague who had used the offensive language. So did Siemens' management, who complained sharply about dismissed workers who, instead of approaching the company's engineers directly with their grievance, went (Irish-style) to government staff who in turn made representation on their behalf. This annoyed the Germans intensely. Herr Heintze pointed out to Professor Rishworth that such behaviour undermined discipline and would not happen if contractors and workmen were of the same nationality.[11] He pleaded with Rishworth to instruct workmen to approach his people directly, little realizing that he was now in the field of anthropology, dealing with a major cultural difference between German contractor and Irish workman which all the instruction in the world would not dissipate.

The urbane Professor Rishworth replied in a delicate magisterial fashion that conceded nothing and said all. He began by gently chiding Siemens management for believing that he could prevent workers from speaking to government engineers and inspectors. But in the interests of harmony and better understanding, he would arrange that any relevant information that he and his team felt would be useful to the contractors would be conveyed to them.

The next paragraph of his letter left no room for misinterpretation:

> It will certainly lead to greater efficiency if the Irish workers are treated sympathetically by the German foremen and engineers, and it is to be regretted that the amicable relations which usually do exist between native employers and their workers on large works are lacking in your case. I am aware of the relations between foremen and workers in Germany, and I know your difficulty with your German foremen trained in this system; Irish workers will not be bullied, but they are easily handled if they are treated with moderation. Of course I realize that you must preserve discipline and that you will find inefficient workers who have to be dismissed, but I would ask you to make an effort to control the language of your German foremen; with their slight knowledge of the English language expressions which to them appear innocuous may be very insulting, and if you now give them a formal warning they will not, on their part, feel they are under a grievance when they are sent back to Germany for offences of this kind.[12]

The final paragraph of his letter dealt with the dismissal of the train driver against whom there was evidence of 'slackness', and because of that there was little Rishworth could do. However, within two weeks he had set in place, with the agreement of the contractors, a formal dismissal procedure for train drivers so as to obviate some of the more peremptory and apparently arbitrary dismissals that had taken place.[13]

Within a few weeks the contractors had derailed this piece of procedural machinery, finding it too limiting and cumbersome for them. This came to light when a train driver from Galway, called Spain, was sacked by the German foreman, Tubbeche. A German driver, da Corta, who had derailed an entire train at Clonlara causing it to fall down an embankment and into a field, was transferred to Blackwater – to replace Spain. The latter was involved in an accident in June when he was working on night shift. He was returning his engine to Blackwater when, as he was passing the stone crusher at Ardnacrusha, he was asked to help a loaded train up an incline – which he did by hooking his engine on from behind. An electric train driven by a German came up from the rear at full speed and collided into Spain's engine, neither train having lights in the dark and foggy conditions. The driver of the electric train had been warned by the points fore-

■ The carpenters at work on the timber shuttering for the penstock frame at the Power Station.

man to be careful of the operation up ahead, but the crash happened anyway. Spain was dismissed without consultation and his discharge paper was marked 'Dangerous'.[14]

Prendergast, the Resident Engineer, was quickly on to his case, calling for his reinstatement until the facts were investigated. He pointed out that Spain had been dismissed through no fault of his own and was replaced by a German driver who had already been dismissed from one building site because of carelessness or incompetence. He added that he considered the dismissal of Spain as a breach of the agreement on procedures for dismissals. He was now demanding an explanation.[15]

The tone and content of Prendergast's letter meant that a 'high-noon' situation had developed between the contractors and the Irish management as regards the dismissal of skilled men. For some weeks the contractors had been in breach of the agreement made in February, dismissing Irish workers when they wished. The Spain case, though, had brought matters to a head and the contractors were not prepared to back off, as a quick reply from Heintze made abundantly clear.

'As our firm pays the wages, the engagement and dismissal of workmen is entrusted solely to the Firm,' he declared. He said that they only signed up to the agreement on dismissal procedures to help Irish management over their political difficulties, and to facilitate the engagement of a further sixty German train drivers. As they had employed only twenty-five of that number (thus not reaching their quota) and were saddled with ninety-eight Irish train drivers (most of whom, according to the contractors, were incompetent), they felt they were not obliged by the agreement and neither had they an obligation to inform the Resident Engineer of dismissals. Heintze insisted that their sectional engineers must have the power to dismiss workmen at any time, particularly train drivers, if order is to be maintained and work kept on schedule. In the case of da Corta, he said (contrary to established evidence with the Irish officials) that he had been transferred from Clonlara to Blackwater as the result of a quarrel with his German foreman. Because of that he had been dismissed at the beginning of July and would be leaving Ireland at the earliest opportunity.[16]

On the face of it, Heintze's letter seems cavalier at best and disingenuous at worst. Of course there has to be an allowance for the fact that the contractors were working through the medium of a foreign language and that phraseology might not be as delicate or felicitous as it could be. But even allowing for that, the literal driving of a train engine through the agreement without informing the other party, coupled with the perverse logic used to justify that move, was astounding. Prescinding altogether from the question as to the efficiency of the Irish train drivers, the German bosses were saying to their Irish counterparts: 'We arrogate to ourselves the right to fire workers and interference by you will only subvert our effective running of the whole operation. We know better.'

Perhaps they did, but many of the Irish engineers were highly experienced in large construction site work and would have resented greatly the German hubris. Rishworth himself had worked in England, Europe and Egypt, and his response to the contractors was swift and to the point: 'I am directed to inform you that if any locomotive driver is dismissed for reasons which in the opinion of the Minister are inadequate, he will not allow him to be replaced by a German driver – and in addition he may, according to the circumstances, direct that the permits of one or more German drivers should be withdrawn.'[17] The contractors were insistent that half of the full complement of 120 train drivers should be German, but by exercising strict control on the granting of permits their number

■ Collecting stones in the shadow of the high-voltage compound.

had only risen to twenty-five. After Rishworth's letter even that number was under threat. Throughout the Scheme the question of permits for German workers was a thorny issue. On the one hand the government was under pressure from trade unions and Irish workers to provide jobs, and on the other, the contractors wanted more and more of their own experienced people engaged in order to accelerate progress and finish within the deadline. At the height of operations they had 830 workmen, not counting their engineers, administrative and accounting staff.[18] They were forever on Rishworth's back complaining about delays in issuing work permits. But the Chief Engineer and the Resident Engineer were very limited in what they could do, working as they were within a political and financial straitjacket, not to speak of the strident local demands for employment. It caused little surprise and some glee in March 1926, when eighteen workmen were not allowed to leave Cobh and were despatched back to Germany because their permits were not in order.[19]

Along with complaining about permits, the contractors indulged in constant and concerted attacks on the ability of the Irish workers, both skilled and unskilled. In fairness they must have had a case, as thousands of the Irish workers were straight in from the fields and were totally lacking in experience of big building sites. However, it must have been tough and tiresome for the Irish authorities to have to entertain a constant barrage for the duration of the works.

The hapless Irish foremen came in for a particularly hard time from the contractors. Earning anything up to £5 for a 50-hour week, their style was a cause for derision and ridicule by the Germans. 'Firstly the foremen have little knowledge of their trade, and secondly, they do not induce the men to work. On the contrary, they loll around with their hands in their pockets near their gangs of workmen, a scene which is familiar to you,' wrote Heintze to Rishworth. He also complained that the concentration of Irish workers was considerably less than the contractors calculated and, as a result, their work suffered. He continued:

'We could under no circumstances employ foremen who do not help us to increase the intensity of work to a normal standard and to incite their gangs of workers by advice and example to a normal standard of work.'[20] The proactive type of foreman the contractors wanted just did not exist among the ranks of the Irish, and of the twenty they had taken on, fifteen had been sacked already and there were more to go. Contractually they were required to employ a certain quota of Irish foremen, but their difficulty was in finding them. This was saying, indirectly of course, that we have to employ more German foremen thereby increasing the leverage for more permits.

One has to question to what degree the contractors were overstating their case, when one notes the instance of C. Coughlan, a foreman with thirty-seven years' experience as a building contractor, employed by Siemens Bauunion, on the repair of the road between Clonlara and O'Brien's Bridge. He was paid 1s. 6d. per hour and told the Resident Engineer that his work could be done by a ganger, and he employed more profitably elsewhere! So much for the inefficiency of the Irish foreman.

But the contractors never gave up. In the summer of 1927 a six-page report landed on Rishworth's desk detailing yet again the catalogue of complaints about Irish workers, skilled and unskilled. Not only was the Irish output lower than the Germans', but there had been serious damage to machinery through carelessness and inexperience – locomotives burnt out and running repairs neglected, a massive dredger destroyed because the stoker forgot to fill the boiler with water, boring plant rendered useless through senseless continual boring, etc., etc:

> 'We feel sure that you appreciate that the mentioned defects are the result of no
> large Civil Engineering work having ever been undertaken in Ireland and the lack
> of development of industry in the country up to the present time. Irish workmen
> are not accustomed to putting any energy into their work which we expect, and it
> is not possible to find skilled men in Ireland of the quality we need'.[21]

Quoting their labour advisor, Joe McGrath, who found it impossible to get suitably qualified foremen for excavation work or supervision of crane operations, they stated that their only option was to bring more men in from Germany and had been so advised by him.

Much of the report had been mentioned in previous correspondence, but there were some novel elements to it. One was the difficulty that emigration posed. Between March 1926 and May 1927, 184 men had left the Shannon Scheme, most of them trained by the contractors. Proof that many of these had emigrated to the US was provided by the number of American firms that had contacted Siemens Bauunion for references. The report commented that these men had left voluntarily and were the best of those employed. It added that the continual retraining of men was a costly exercise. It is understandable that the contractors were frustrated by this pattern of social life in Ireland, but at the time emigration was endemic, with 43 per cent of Irish-born men and women living abroad, over one million of them living in America.[22]

A second interesting element in the report was the mention that many Irish workers, skilled and unskilled, frequently did not turn up for work on the pretext that they were ill. The contractors, though, were of the opinion that they simply wanted to take a holiday from time to time. They backed this up with a survey showing the average hours worked by German and Irish skilled workmen. Their own people worked an average of 64.5 hours a week over a 23-week period, whereas the Irish worked 55.5 hours a week during the same period, a difference of 14 per cent. In addition to hours worked, they also calculated absenteeism during a five-week period and found that 51 Germans were absent for 30 days, while 109 Irishmen were absent for 222 days in the same period. Translated, this meant that each German was absent 0.6 days during five weeks, while each Irishman was missing for two days.[23]

The report was pretty conclusive, but just in case it was not enough, Siemens Bauunion followed it up with another report conducted by their colleagues in Siemens Schuckertwerke who had just conducted a performance trial of Irish and German gangs on the transmission line between Dublin and Kildare. The result of the trial should have been a foregone conclusion as the Irish workers had only had a few weeks training, while each of the Germans had engaged in this work for five or six years. The contractors were quite positive about the outcome of the test and felt that the Irish workers had acquitted themselves well, relatively speaking.[24] The sister company on the civil side used this report as further ammunition in its war of attrition with Irish management, which, in a way, was a measure of German frustration with the situation.

While the dogfight continued at high altitude, the daily tasks of fire brigading had to go on at the various sites. From correspondence it would appear that the Irish management at local level were on the back foot most of the time, investigating incidents after they had happened and struggling to right wrongs that should never have taken place in the first place.

For example, a train-driver named Flaherty from Dublin was dismissed. When he got his discharge ticket translated from German he discovered that he had been let go because he did not want to keep his engine clean! Hard to believe in a place where there was muck up to the waist.

An interesting case took place in March 1928 when a train-driver called John Gilchrist from Charleville went down with influenza. He had been a very satisfactory worker for two and a half years, operating mostly around the workshops and the temporary Power Station loading and unloading material. Dr McSwiney saw him when he got sick and sent him home for a few days, instructing him to see his own doctor in Cork. He returned to work after two weeks and again was checked by the camp doctor who issued him with a sick certificate. When he presented himself for work he was told by his German foreman, Faht, that there was no work for him as his machine had now been given to a German, Zarcke. Gilchrist spoke to Stolberg, the Siemens engineer in charge of the Ardnacrusha site, who told him he could do nothing for him and that he could go labouring if he wished. This would mean a considerable drop in wages for him, as he was earning £3 2s. 6d. a week as a crane-driver.

It transpired that the German driver was now earning £4 10s. for 50 hours and 1s. 9d. per hour for overtime. It also became obvious that Zarcke was incapable of driving the crane. A few days previously he had broken the gears and men complained that they had to wait for over an hour in the morning before he succeeded in starting the engine. They also claimed that he was dangerous to work with. Gilchrist was an experienced driver and could start the crane in three minutes.[25]

The Resident Engineer took the matter up with Siemens and it quickly became clear that the contractors were simply trying to find work for their man instead of sending him back to Germany.[26] The fact that he had discommoded an Irish worker did not seem to matter. After an official of the contractors, Herr von Brewick, checked the case, correspondence on it suddenly ceased, which would indicate that the Irish officials had got their way and that Gilchrist was reinstated as crane-driver.

A number of other documented cases of not only unfair dismissals, but 'extremely outrageous' ones, made for difficult relations between Irish and German management. In some cases there was a strong suspicion that workers were 'set up' to be got rid of, particularly if permits had been issued to German workers to come on site. One of those was of a boatman called Fitzpatrick who worked at the weir.[27] Another was a carpenter named King (an ex-national army man)

■ Strengthening the foundations of Killaloe Bridge.

who was either struck on the face or, on whom there was an attempted strike, with a wire rope by a German carpenter at the weir. There was an exchange of words and King was discharged but was later reinstated. The Irish official who investigated this and other cases of dismissals commented in his report: 'I know that the least dispute between a German and an Irish workman is sufficient to get the Irishman discharged, either then and there, or later, no matter whose fault it is.'[28] This view was confirmed repeatedly in various cases, but none more so than in that of James Langton, a storekeeper, who was dismissed even though a German fitter had threatened to smash his head in with a lump hammer for the flimsiest of reasons.[29]

By the end of 1928 the major part of the excavating and concreting work on the Scheme had been completed. Men began to anticipate that in a short time they would be let go. The foremen, particularly, began to jockey for position, as some of them would have to be retained until completion of the works. The Resident Engineer picked up on this feeling of unease among the men and decided to pre-empt a problem that might arise rather than let it happen. He wrote to Rishworth and to his colleagues:

… I am of the opinion and always was that it is quite unnecessary to have German foremen in charge of small gangs in excavation and concrete work. There are a number of such German foremen on the work and I suggest that we should insist that the men be sent back to Germany in preference to Irishmen being dismissed. I would further suggest that the contractors should be warned before any actual case of dismissal occurs'.[30]

In March 1929 the wholesale discharge of tradesmen began because of 'slackness of work'. Carpenters were the first to go. Train-drivers and stokers were next, followed by blacksmiths, fitters and any other tradesmen who had now become redundant.[31]

By early June the problem that Prendergast had identified in March as a possibility had developed for real. Two general foremen on the headrace, Purcell and Casey, had approached him with the news that the former had been told by his German engineer, Harbaur, that he would shortly be let go; the latter was told that he would also get his notice in the not-too-distant future. Both men felt that they were entitled to preferential treatment as they had stood by the contractor during the strike at the beginning of the Scheme and that their work as foremen had been satisfactory throughout. They realized fully that their work was co-terminus with the Scheme, but their complaint was that they were on the point of being dismissed while there was still work to do and Germans were being retained to carry on the work of general foremen, duties which they could happily and efficiently fulfil.

Prendergast wrote to Siemens challenging the prospective dismissal of these two men, while retaining their own men to do work that Casey and Purcell could adequately do:

> I am not writing this letter with a special view to your examining the specific cases of Purcell and Casey but more particularly to have the general question of dismissals of Irish foremen raised and a clear understanding of the procedure you intend to adopt as the main construction work nears completion.[32]

There was quite an amount of subsidiary work to be done, such as trimming of the canal banks, soiling, clearing of land, road and bank maintenance, fencing, etc. – all of which could be supervized by Irish foremen. Prendergast left the contractors with little room for manoeuvre on this one: 'I should like to have your assurance that Irish foremen will be retained for this work and that so long as German foremen are retained, Irish foremen will also be retained unless it can be shown that they are incompetent.'[33]

The National Ex-Servicemen's League who had played an important part in the strike monitored the situation and wrote to Prendergast:

> There is no need to point out to you that our men as engaged are doing the identical work as general foremen, for which the German is receiving at least 8 pence per hour more ....We trust then that you will be kind enough to use your good offices on our behalf as we hold that being in at the 'break' we should also be in at the 'kill'.[34]

Meanwhile, tradesmen continued to be discharged, but in July as the Scheme was being opened four Irish plasterers were let go at Ardnacrusha. They were quickly reinstated when Rishworth warned the contractors that there would be trouble with the other Irish workers as eleven German plasterers were still being employed there.[35]

By December Casey and many of the other Irish foremen were unemployed. A skeleton staff of Germans continued to operate, disengaging from sites such as Longpavement. Their continued presence rankled with Casey and his colleagues in the Ex-Servicemen's League. He wrote to Prendergast:

> I consider it unfair to me that a German foreman should reap the reward of my and my equals' efforts. I was wounded in the struggle for Irish freedom and was also wounded in the National Army, so I would feel grateful if you would kindly have my case carefully investigated and inform me at your earliest of the result. Thanking you for past favours ...[36]

The Resident Engineer sent a copy of Casey's letter to Siemens Schuckertwerke with a covering note saying he had recently met Casey and told him that in view of the near completion of the works it was unreasonable of him to expect preferential treatment at this stage of the work.[37] Perhaps – but not really, when it is clear from the McGilligan papers that people were being rewarded for services rendered. Joe McGrath had contacted the Minister about a number of individuals who had served them well, such as Commandant Stapleton who was in charge of the camps at Ardnacrusha, and Tim Murphy, the stevedore from Limerick. Even though McGilligan was away in Geneva he had his officials contact James Fay who was shortly to be appointed to the Electricity Supply Board to ensure that the latter would look after the two men with jobs in the ESB.[38] It was not for Prendergast to know this, of course, but it is interesting to note how political patronage was being exercised by those who had the power to do so.

How things had changed! There was a time when Prendergast would have fought Casey's corner and challenged the contractors as to how they were operating. Not any more. It was now Christmas 1929 and, at last, in good seasonal fashion everybody was singing off the same hymn sheet. The delicate problems of human resources and administration were thankfully a thing of the past. The 5000-plus Irish workers had returned to the four corners of the country and their huts, canteens and makeshift villages were dismantled. Most of the heavy machinery had been shipped back to Germany and nearly all of the German workers had gone too. Ardnacrusha now stood as a monument to international cooperation in spite of the racial tensions, the contrasting management styles and the varying work-rates of those involved. It also stood as a tribute to the diplomatic, political and negotiating skills of German and Irish executives. For a country that never really had an industrial revolution (apart from the northeast), working on the Shannon Scheme was the closest that thousands of Irish workers of that generation came to experiencing what the industrial revolution may have been like. In any event, File SS13929 is a great measure of the distance travelled by the science of human resources in this country in seventy-five years.

5 | Living Rough

The building of the Shannon Scheme affected the public consciousness in varying degrees. The prominence the project enjoyed in 1925 because of a major industrial dispute, which for a time seemed to threaten the whole prospect of the Scheme, receded somewhat at the beginning of 1926 when the strikers were out-manoeuvred. The occasional progress report in the newspapers was all that was to be seen of the undertaking during the following spring and early summer. But in late June the public mind again focused on Ardnacrusha as newspapers called for a special enquiry, alleging that workers were living in barns, stables and pigsties.

The ferment of ingredients for this latest notoriety had been bubbling away merrily for months. Employment at the works had risen rapidly to over 2000 since major construction had begun in September 1925.[1] Maximum sleeping accommodation for the Irish workers, skilled and unskilled, at Ardnacrusha camp was 720 and that was only achieved in 1928 when employment rose to as high as 5000.[2]

In 1926, therefore, approximately 1300 had to find accommodation either in Limerick City or in the neighbourhood of the works. Besides, scores of men walked to Ardnacrusha from all parts of Ireland looking for a job at the going rate of 32s. a week. Some were sent there by their local Labour Exchange, many others went there full of hope and, oft-times, very little else.[3] There was no dole in those days; neither was the prospect of employment elsewhere very bright.

■ Top: The camps could only accommodate 700 of the 5000 workers. Above: For those who would afford it the camp accommodation was clean but basic with up to 26 in a hut.

For less that a quarter of the 2000-odd employed on the Shannon Scheme in June 1926, free sleeping accommodation was available in the site camp in one of the timber huts provided by the contractors.[4] In the neighbourhood, lodgings could cost anything from 2s. to £1 a week.[5] Meals could be bought in the camp canteen at the reduced rate of 11s. 8d. a week.[6] A pint of porter then cost 6 pence. Health and unemployment insurance deductions amounted to 1s. 1d. a week.[7] A visit to Limerick's Coliseum or Grand Central to see the latest silent Tom Mix or Gary Cooper film would cost 1s. or 1s. 6d.[8] So, for those lucky enough to get jobs the financial pressure of having to provide their own accommodation was considerable, particularly if they were married and had hoped to send home something to their wives and children. The less fortunate were confronted with two options: the long dusty road home or just hanging about in the hope that the following morning might bring a job for them. In the meantime, however, they had to find somewhere to sleep. One of the most obvious places for those without work to look for accommodation was in the City Home in Limerick. During April, May and the beginning of June 1926, an average of ten people unable to find work on the Shannon Scheme were admitted each night. Towards the end of June the number seeking admission had dropped to five a night. But then the City Health Board refused admittance to all those coming from outside the Borough of Limerick.[9]

The decision of the City Health Board to refuse accommodation to all except Limerick townsmen was related to another question that had been occupying the minds of Clare County Board of Health and Limerick City Home and Hospital Committee for some months, namely: who was responsible for the medical care of the Shannon Scheme workers?[10] The German contractors refused to hold themselves responsible, stating that their liability ceased when they complied with the terms of the National Health and Unemployment Acts. Clare County Board of Health held that even though the Shannon Scheme was situated in Clare, its hospitals could not cope. It argued that Scarriff District Hospital, catering for east, central and north Clare with only thirty beds, was already inadequate for current needs, and that although the County Hospital, being the County Home, held a certain number of beds for the poor of the county, it had no extra beds available for the Shannon Scheme workers. Limerick Board of Health argued in similar vein that it was incapable of meeting the needs of the Shannon Scheme and that the Minister for Industry and Commerce should make the contractors shoulder their responsibilities.[11] In fact, there was a medical officer at Ardnacrusha, Dr Myles McSwiney, who had a few beds for the injured and sick, but nothing to cope with the kind and volume of injury that invariably happens when you have thousands of unskilled men working under pressure with heavy machinery and explosives.

It was on Thursday, 24 June, at one of its monthly meetings, that the Clare County Board of Health first heard of the Shannon Scheme workers sleeping

■ The general store at Ardnacrusha camp where many other shops sprang up.

in pigsties and stables.[12] The Board had just listened to a report from its Secretary, John Quin, on a conference between the Limerick and Clare County Boards and the Department of Local Government on the subject of treating the sick and injured of the Shannon Scheme. The conference had requested the Minister for Local Government to introduce in the Dáil a bill that would allow the transfer to their homes of workers from other counties or countries in the event of their being destitute, sick or injured. The representative of the Department had told the conference that Clare and Limerick County Boards of Health could refuse relief to any man coming from outside their respective areas. The members of the Clare Board warmly welcomed the report on the conference. As a postscript to its meeting, the Home Assistance Officer (HAO) for Ardnacrusha was asked for a report on the latest position in the area. HAO Mullane stated that a lot of men had billeted themselves in stables, cowsheds and barns. There was, he said, no hut accommodation for many of them. He reported that a man and his wife had taken possession of a pigsty attached to a labourer's cottage, and there were no less than twelve to fourteen in a stable at Blackwater in O'Grady's yard. Some of these men were unemployed, some were not. The HAO informed the Board that about 400 workers were being laid off within a few days and he wanted to know what he would do with those who remained in his district and became destitute. There was, he said, a constant procession through eight miles of his district by men looking for work or dismissed or going home after failing to get work.[13]

A spirited debate followed. Councillor Pat McMahon from Blackwater corroborated the statement of the Home assistance officer that people were living in pigsties and that some men (and their wives) after being dismissed from the Shannon Scheme remained on in the district. 'Surely to God,' he exclaimed, 'we are not going to let them die with the hunger.' Councillor Halpin was of the opinion that the government should shoulder its responsibility to save people from living like 'mere swine'. Other members of the Board echoed this view. Councillor Halpin also cited the case of a man lying in a pigsty with straw for a bed whilst he was suffering from double pneumonia. Councillor Crowe asked, 'Are we to feed the hungry of every county in Ireland?' McMahon replied in typical generous fashion, 'We would and divide our last penny with the poor, but if we have not a penny, what then?' The Board resolved, however, that it would abide by the principle of the inter-county conference and grant relief only to Claremen.

It also ordered a comprehensive report on conditions in and about Ardnacrusha from the Medical Inspector of Health for the area, Dr G.H. Enright.

Two days after the County Board of Health meeting, the national press had picked up the story of housing conditions on the Shannon Scheme.[14] The *Irish Independent* recorded the Home Assistance Officer's statements at the Board of Health meeting in its editorial page under the heading 'A Very Amazing Story'. Three days later the story had been moved to the world news page and a call was made for a special inquiry and for an early remedy for the inadequate housing conditions of the Shannon Scheme workers. The newspaper had also made its own investigation and published an exclusive report that corroborated the statements of Home Assistance Officer Mullane. Referring to the pigsty attached to the labourer's cottage, the special report stated:

> The latter accommodates a husband, a wife and two children. Some of the places in which men are sleeping are not at all fit for human beings. There was one place referred to by the Home Assistance Officer. It was merely an outoffice; it might have housed horses or cattle. The beds consisted of old hay, thrown on the floor, with no suggestion of bed clothing. One of the heaps of hay was semi-covered with an old sack. This was the very place in which 15 men slept up to a short time ago. The number has now dwindled down to eight and those men are paying rent

for the privilege of accommodation. They are being charged 2s a week for the shelter of the roof and the bed of hay self-provided. Other places in the locality, outhouses and so on have been improvised and are quite alright, but the place to which specific reference has been made ought not to be allowed to exist. The Camp Commandant at Ardnacrusha, Mr W.J. Stapleton, when interviewed on the matter, stated that it was possible that there were men living in outhouses, but they were not employed on the Scheme and that it was not because of lack of accommodation. 'This morning there were only two men on my waiting list for beds,' he said. He further explained to me that some months ago before the huts were completed there was not sufficient accommodation, and the men were obliged to fend for themselves and do the best they could; but now they were able to cater for them. He went on to say that when hut accommodation was not available, the fact was stated to the men before they were employed, and as a result of the rapidly approaching completion of the temporary works at Ardnacrusha, between 200 and 300 men had been disemployed during the past month, and whilst an effort was being made to give them at least one night's accommodation after dismissal, they could not be held responsible for them afterwards. A further matter which he pointed out was that men were arriving at the works daily from all parts of the country, oftimes penniless, seeking to find employment and unable to find any. It was quite conceivable that those men in numerous instances were obliged to sleep out.

The special report also called for an impartial inquiry into housing conditions for the workers.[15]

One would imagine that at that stage the unions would have become seriously involved in a campaign to improve living conditions and would have used it as a platform to raise the general lot of Irish workers on the Shannon Scheme. Unfortunately there is little evidence to suggest either. A meeting had been held in Ardnacrusha on Sunday, 17 April, and was addressed by Cathal O'Shannon and Paddy Hogan TD in support of a move to organize the unskilled workers on the Scheme and to get them into the ITGWU. The two speakers stressed the need for organization to improve wages and living conditions; a resolution to this effect was adopted.[16] That was the last reported union meeting on the Shannon Scheme and the last indication of any union activity for the remainder of the contract. The union subscription was 6 pence a week.[17]

Even *The Voice of Labour*, which had played such an important role in the Shannon Scheme strike six months earlier by its uncompromising stand and reportage, showed only a short-lived interest. On 3 July it gave front-page coverage to the situation at Ardnacrusha.[18] Under the heading 'Not Fit For Human Beings',

■ A typical German wooden house in the neighbourhood of the works near Blackwater.

it stated that the revelations of the correspondent of the *Irish Independent* may have been amazing to the capitalist press, but not to *The Voice of Labour*. Another edition carried an angry letter signed by 'Man from Nowhere'.[19] The writer stated that he had seen worse slavery in Mexico. He described the working conditions – the bread, tea and margarine for breakfast; the extra cup of tea for two pence and the extra slice of bread for one penny. He quoted a worker as saying: 'We are working for our chuck, and slowly starving to death on it'. Having condemned the bullying German and Irish bosses the writer declared: 'What is the labour movement, the republican movement and the Church doing to expose and do away with this infamy? I would warn all three to be up and doing, particularly the latter, as their silence is roundly condemned by the workers in the huts after their day of slavery.'

That letter was followed by another signed by 'A worker on the job'.

> The conditions on the Shannon Scheme cannot, and I will venture to say, have no parallel in Europe at present. Take, for instance, the wages of 8½ pence an hour with broken time – and God knows for some weeks we get lots of that. Most of the men are in debt every weekend, and I may add if they were to eat enough of good plain food as navvies shovelling concrete, or up to their knees in water

attending the diggers down below, the wages of even the full week which amount to 34s, would not be sufficient to buy sustaining food for a hard-working navvy. I have seen my countrymen housed in the cowsheds of England and Scotland years ago. There, at least, the sheds were fairly warm and sanitary compared to the sheds of the farmers of Blackwater where my countrymen are housed now and pay 18s a week for a half-roofed shed. If William Cosgrave or Kevin O'Higgins were forced to spend a week or two living in O'Grady's cowsheds on the Blackwater, I wonder would they rebel against those conditions, or would adversity make them as slavish as their countrymen, whom they have degraded by forcing those conditions on them? Let your Labour members ask Mr McGilligan why are three or four hundred men per week sent from labour exchanges to the job, and a similar number sacked after from one to six weeks' work and left to drift back home, and sacked for no cause. Ask McGilligan why it was a man was crushed to death between two wagons when the Irish foreman told him not to go on the wagon as it is dangerous. The German told him to get up or go. His son gave evidence to this effect and was immediately dismissed. Ask him why the Camp Commandant Stapleton will pull down a notice the men put up, and tell them 'they must not talk of a union here'. Ask him why it is that when the men put up a lean-to for the purpose of cooking, because their wages won't buy them sufficient food in the canteen, it's pulled down....

Last week the workers had one holiday and two wet days. No money to buy food this week. Yet another holiday today, Thursday! We would advise the workers to get busy and give them the following advice:

> Quit hollering out for martyrs
> To be slaughtered in your sight;
> Get off your knees, you lobsters,
> And learn to think and fight![20]

The apparent lack of involvement by the unions in the issue of living conditions, following so rapidly as it did on the heels of the strike that had held up work on the Scheme for months, may have been due to the fact that the unions were at this point demoralized. In fact the ITGWU decided not to set up a branch at Ardnacrusha or at any of the building sites. Joe McGrath had out-flanked and out-played the unions during the strike by using ex-servicemen's labour.[21] The unions never recovered from this setback, and McGrath's ruthlessness and tight-ship style favoured the employment of ex-Free State army men, of whom the camp commandant was a captain, and excluded would-be troublemakers and union organizers. The contractors also, reportedly, indulged in counter-moves by encouraging the formation of bogus unions in the Ardnacrusha camp, setting

■ Otto Rampf (right), an excavator driver in the tailrace, enjoys a drink with two German colleagues. The sign reads 'in good times and in bad we will remember you'.

up an effective camp CID, and by employing a 'heavy gang' to keep law and order.[22]

Housing conditions in Ardnacrusha surfaced in the Dáil on 1 July. Paddy McGilligan, the Minister for Industry and Commerce, was asked by Deputy Lyons (Longford-Westmeath) whether reports that appeared in the press as to the manner in which men worked on the Shannon Scheme were correct? The Minister replied: 'You may take it prima facie that the press is inaccurate.'[23] But the Munster deputies were not to be as easily dismissed. Later in the afternoon Deputy Paddy Hogan (in whose constituency the Shannon Scheme was based) gave notice that on the motion for the adjournment he would raise the question of the entire lack of accommodation for the care of the workmen on the Scheme when they became sick and injured. McGilligan knew what Hogan was about and tried to head him off by stating that there was accommodation in the camp for 140 more men than were there then. But Hogan was not to be dissuaded and repeated his intention of raising the question of accommodation during the adjournment debate.

Hogan himself made the opening speech. He dealt first of all with the question of accommodation for sick and injured workers:

> A worker from Co. Mayo who has been unemployed, say, for a period of twelve months, or, perhaps two years, and he has therefore lapsed in National Health

Insurance. He has nothing to his credit in the records of the National Health Insurance Society. He gets work on the Shannon Scheme and after two or three months he becomes ill or injured. He has eight or twelve National Health Insurance stamps, probably, to his credit and he has no claim upon the funds of any National Health Insurance Society. He has to fall back on the arrangements made by the local health authorities of the district, while the local health authorities have only made arrangements to meet normal conditions.[24]

Hogan argued that since the Shannon Scheme was not a normal type of work and since the local health authorities were unable to cope with the situation, the Minister should make special provision for the workers. He then got on to the question of general accommodation: 'I would like to say that I am not very much interested in the opinions expressed by the Minister in the press report – the opinion expressed by him this afternoon in reply to questions addressed to him by Deputy Lyons.'[25] He quoted from the report of Home Assistance Officer Mullane to the Clare County Board of Health and from the minutes of the County Board meeting where HAO Mullane was complimented by the Chairman of the County Board 'on his report as to the extraordinary circumstances in which people were living – in out-houses, pig-sties and barns and lying on straw'. From information he had received from public officials in the Ardnacrusha area he

■ Not doing too badly on German emergency rations. Rampf is seen on the right.

supported the veracity of the Home Assistance Officer's report and called on the Minister to state what arrangements he had made regarding the sick and injured and in regard to accommodation in general. Deputy Clancy from Limerick supported Hogan:

> If something is not done very soon as regards the sanitary arrangements at Ardnacrusha, I greatly fear that an epidemic will break out that will be more serious than the 'flu epidemic we had in this country in 1918 and 1919. I was there in the months of April and May, and I must say that the air was anything but pleasant … If that condition of affairs is allowed to continue, I think that the people of Limerick and Clare will very soon have to turn their noses towards the Government and appeal to have something done.[26]

The Minister for Industry and commerce replied to both speakers with what could be described as a typical McGilligan broadside: 'I put in direct negation to what Deputy Clancy has said that there is a medical officer in the camp, and that the sanitary arrangements at Ardnacrusha are as sound as they could be on any works. As far as these arrangements are concerned, there is not the slightest fear of any attack or any epidemic such as the Deputy has outlined here tonight.'[27] As for Paddy Hogan's arguments, the Minister did not deal with them directly but made a vicious personal attack on HAO Mullane. 'We have the statement made by his Home Assistance Officer who strangely enough put in some very hard work indeed for a fortnight before he issued his report trying to get a son or a relative employed on the scheme under these filthy conditions that he has described. The bad sanitary arrangements and the other matters that he refers to would appear to have only been discovered when he could not procure a post for his relative under the scheme.'[28] The Minister then insisted that it would be outrageous to expect the contractors to provide hospital beds for all the sick and injured. There were a few beds in the camp for accident cases and it was up to the Clare and Limerick authorities to look after the rest, he said.

The debate went on until after 10pm, with McGilligan defending stoutly and not giving an inch. In conclusion, Deputy T. Murphy (West Cork) put it to him: 'Has the Minister any statement to make with regard to the report that appeared in the *Irish Independent*, on the conditions, from a journalist who was sent specially to investigate matters?'[29] McGilligan replied, 'I saw the statement there of the Commandant of the camp which, I think, answered the journalist's statement perfectly.'[30] The Dáil adjourned at 10.25 pm.

On the following day *The Irish Times* reported the Dáil debate and the next issue of *The Limerick Leader* also carried a parliamentary report, but it was left to *The Clare Champion* to defend its countrymen in the face of the Minister's attack.[31] McGilligan had accompanied President Cosgrave to a Cumann na nGaedheal convention in Limerick City Hall on 10 July. During their visit they took time off to tour the works at Ardnacrusha and to inspect the accommodation. At the convention though, the Minister was attacked for his statements in the Dáil on living conditions at the Shannon Scheme and his arguments were flatly rebutted.[32] Seemingly the experience at the convention mollified McGilligan somewhat. The next issue of *The Clare Champion* reported:

> During his recent visit to Limerick the Minister for Industry and Commerce seems to have been convinced that a number of workers employed on the Shannon Scheme were living in outhouses in the neighbourhood of Ardnacrusha under insanitary conditions, and he admitted that the Boards of Health of Limerick and Clare in drawing attention to them were not activated by unworthy motives. Mr McGilligan would have done the gracious act if he had also withdrawn his suggestion that Home Assistance Officer Mullane reported the conditions merely because a relative had not got a job on the works. Mr Mullane reported the conditions in the ordinary course of his duty. In his official capacity, Mr. Mullane would have been blame-worthy if he had allowed men to crowd into barns and stables without remark. There was no suggestion that the Shannon Scheme contractors were responsible for the actions of the men, and no slur was cast upon the manner in which the works are being conducted. But when men herd in stables and pigsties, there is an obvious menace to the public health, and if timely measures were not taken to deal with it the consequences might be very grave.[33]

The Medical Inspector of Health for Ardnacrusha, Dr G. Enright, visited the area on 5 July. His report on accommodation at the Shannon Scheme was pre-sented by him during the same week to a meeting of the Clare County Board of Health.[34] He found that seven houses in which workers lodged – Henneberry, Haskins, Lavery, Greensmith, Leahy, Ryan and Hogan – were found to be satis-factory. An eighth house was somewhat congested. In a ruined building on Hartigan's land, in which eleven men lived and slept on shavings, Dr Enright found conditions unsatisfactory. The place was not habitable and should have been closed. In William O'Grady's yard there were eight or ten occupants living under insanitary conditions. In a sty attached to Keegan's labourer's cottage there was a husband, a wife and two children living under insanitary conditions. All

■ The dining-room of Doonass House, near Clonlara, where some German engineers and Dr McSwiney lived.

of the men living in those places, with one exception, were working on the Shannon Scheme. Dr Enright also reported that a number of independent traders had opened shops in the district. They were all very clean inside, but the outside sanitary arrangements were defective. As to hut accommodation at the works, Dr Enright said the average number in each hut was twenty-six.

A discussion of the report followed. Mr Kerin, Chairman of the Board, said that Dr Enright's report repudiated the Minister's remark in the Dáil and corroborated the report of Home Assistance Officer Mullane. Mr. P. McMahon said that the man in the pig sty was earning 1s. an hour and so could not afford to stay in the huts himself and pay for his wife and children's accommodation elsewhere. An order was made to close the ruin on Hartigan's land and the stable at O'Grady's yard, while a directive was also given that the sty attached to the labourer's cottage should not be sublet. The Chairman closed the meeting with the general observation that 'the Poor Law was never intended to meet an emergency of this kind, and the Government and the contractors should shoulder their responsibilities'.[35]

The order to close the ruins on Thomas Hartigan's land could be considered an ironic side to the whole episode. In November 1925 the government acquired Hartigan's farm at Ardnacrusha for use as part of the Shannon Scheme development. Hartigan, of Quinville House, Parteen, like most of the other farmers and landowners in the district, had not yet been compensated for his loss of land. So when the order came from the Clare County Board of Health to close up the buildings on the farm, he reacted swiftly and angrily. He wrote to the Board: 'The Germans took that farm from me last November. They took the stones out of the doors and windows and converted them to their own use. It is bad enough to have my own land gone and no sign of getting paid for it without being compelled to look after it.'[36] In fact, it is now clear from the McGilligan papers that after nine months waiting, Hartigan's claim for compensation had passed that point of the complex government process of acquiring land where ministerial approval had just been given to department officials to take the lands permanently.[37]

While Hartigan may have complained about the government taking his land and the County Board of Health asking him to maintain it, his complaint was minor in comparison with the anomaly, as pointed out by a Justice in the Civil Bill Court in Limerick in 1928, where one government department was taking land without paying compensation and another government department was suing for land annuities![38]

The constant pressure from the officers of the Clare County Board of Health seems to have eliminated the worst of the accommodation abuses at Ardnacrusha, temporarily at least. Where the occupants of the stables and pigsties went for accommodation one is not quite sure. But perhaps a clue is provided by *The Limerick Leader* of 6 October. The Clare County Board has been discussing of late the illegal erection of thirteen huts on plots tenanted by locals.[39] When called on by the County Board to explain why they had allowed huts to be erected on public property, one tenant wrote: 'I wish to inform you that my plot is of no value owing to the work in progress on the Shannon Scheme being carried out in the vicinity. The only way out of the situation was to sublet it to a few friends who have given me a small fee for the use of the huts for a few months. Only for so doing myself and my family would be on the verge of starvation …'[40] Another tenant explained his position: 'I wish to inform you that my plot was of no value to me owing to the work carried on here. I could sow nothing in the plot as it was encroached upon by every form of trespass. The only thing left to me to do was to oblige a few who helped me by giving me a small item for the place for a few huts for a few months.'[41] Members of the County Board of Health noted the correspondence but they also noted that employment on the Scheme was topping 3000, and considering what happened as regards accommodation when numbers employed were considerably lower, they decided to take no action against the tenants and to leave the illegal huts.[42]

Christmas 1927 at Ardnacrusha was quiet, and the locals reportedly pined for the New Year which would signal the Connemaramen's return with a new consignment of poitín.

The New Year brought the Connemaramen back to the Shannon Scheme, but it also brought echoes of the housing scandal of 1926, and worse. The first week

■ President Cosgrave winches up the first high-voltage tower at the invitation of Carl Freidrich von Siemens in February 1927.

of February saw *The Clare Champion* with the front-page headline, 'Crowds in Stable – Shocking Conditions at Ardnacrusha.' Dr G. Enright, Medical Officer for the district, had reported to the Clare County Board of Health that housing problems had again arisen in the Shannon Scheme area. O'Grady's yard in Blackwater was singled out by the Medical Officer as the main place requiring the attention of the County Engineer's office. In the past year workers had occupied O'Grady's yard for a time before officers of the Board of Health evacuated them. This time, workers, their wives and children constituted the 'households', which swarmed in the stables and outhouses of O'Grady's yard.[43]

Dr Enright listed the following families being present:

> Mr. & Mrs. Dennehy & 1 child;
> Mrs. & Mrs. Sheehan & 3 children;
> Mr. & Mrs. Greensmith & 5 children;
> Mr. & Mrs. O'Halloran & 3 children;
> Mrs. Rosethorne and brother;

Mr. & Mrs. Lee & 2 children;

Mr. & Mrs. Leahy & 4 children;

Mr. & Mrs. O'Dwyer & 6 children;

Mr. & Mrs. Spain;

Mr. & Mrs. Hogan & 5 children;

Mr. & Mrs. Moloney & 2 children;

Mr. & Mrs. Haskins & 2 children;

Mr. & Mrs. Henneberry & 1 child;

Mr. & Mrs. O'Callaghan;

Mr. & Mrs. MacMahon & 2 children;

Mr. & Mrs. Lavery & 3 children;

Mr. & Mrs. Ward & 4 children;

Mr. & Mrs. Leahy & 3 children;

Mr. & Mrs. Higgins & 3 children;

Mr. & Mrs. Shanahan & 2 children and

Mr. & Mrs. Moore & 2 children.

The total number of people in the yard came to ninety-four. In the Hogan family mentioned above, Dr Enright reported that the husband had been moved to hospital suffering from typhoid fever, having been admitted on the certificate of Dr M. McSwiney. The case of typhoid shows, said Dr Enright, that 'the drainage requires immediate attention whilst the list of families resident in the place makes it obvious that the area is very congested'.[44] It is believed that the occupants of the stables and outhouses were being charged £1 a week.

The next meeting of the County Board of Health took place on 9 February 1928. A letter from the Department of Local Government was read acknowledging receipt of Dr Enright's report. The Department 'presumed that the Board of Health will obtain further information from the Medical Officer of Health as to the character of the housing accommodation, whether the congestion is due to circumstances of a permanent or temporary nature and what remedial measures are available, and will then proceed to consider the advisability of dealing with the situation so disclosed under the powers conferred on them by the Public Health Acts'.[45] This bureaucratic jargon returned the ball to the County Board's court and effectively forced the Board to close down O'Grady's yard once again.

Nothing more was heard of housing conditions on the Shannon Scheme during its remaining construction period, even though employment rose at one stage to over 5000, thereby putting extreme pressure on whatever accommodation was available.[46] In general, like the bitterly-fought issue of labourers' wages that threatened the prospect of the Shannon Scheme in 1925, the issue of living conditions for the workers, as a platform for change in the fledgling republic, was gradually vacated by public representatives and politicians. Organized labour had passed up another opportunity to put its own stamp on the biggest industrial undertaking since independence.

The completion of the Shannon Scheme was due in no small measure to the brilliance and determination of Paddy McGilligan, the Minister for Industry and Commerce. There is no doubt that he could have put more pressure on the German contractors to improve the lot of Irish workers. He argued publicly, at all times, that it would be unfair and unreasonable to expect the Germans to carry any more responsibility than they were already doing. But the Germans, in fact, cannot really be faulted because, as is now well known, they were quite prepared to pay more and to raise the standard of accommodation. McGilligan's inflexibility in the face of protracted disputes and oftimes severe pressure inevitably made him the object of criticism and satire for many a pen. At a time when living conditions and workers' accommodation were in dispute the following was a contemporary example of one poet's efforts:

> I am McGilligan, McGilligan, McGilligan!
> Never shall you see such wonderful skill again.
> I've got a plan on
> Down where the Shannon
> Rolls to the foaming sea.
> Jealous folks say it is a bit of a gamble
> But we know better, don't we?
> For there's Siemens Schuckert and Gordon Campbell
> Joe McGrath and McLaughlin and me!
> So now come all ye unemployed,
> You surely should be overjoyed,
> At the finest job that was ever seen
> Thirty-two bob and a cheap canteen;
> We want every man on,
> So come to the Shannon
> Come down to the Shannon with me.

I'm McGilligan, McGilligan, McGilligan!
See me bend resistance to my will again
From Derry I came
To the Halls of Fame
Like a blast from the North
My decree has gone forth,
and as I have told you before
There is thirty-two bob
For each man on the job
And never a half-penny more …
They say it's outrageous,
But I say it's courageous,
To pay to a navvy agricultural wages; …
Anyway there's a job
At thirty-two bob
Ready to begin
With the food thrown in,
So don't mind the ban on
Come on to the Shannon
Come down to the Shannon with me.

Chorus
Dot vos McGilligan, McGilligan, McGilligan,
He will our empty pouches fill again;
Ja! Ja! Ja!
Ja! Ja! Ja!
Oui! Oui! Oui!
C'est comme ça.
Dot vos von
Great big man
Ja!
Ja!
JA!
McGilligan.[47]

6 | Trouble by the Mile

Standing on the footpath of the tailrace bridge, looking upstream towards the Power House, one can only marvel at what the contractors achieved in making this mile of river. All looks settled now, even natural, as the canal lazes its way towards the Shannon farther down. The scene is so lovely it could easily make one forget the enormous technical difficulties in its construction, particularly as these were compounded by complex and searching human problems, all of them costing a lot of money.

On immediate inspection, as one looks down from the bridge, one can readily spot signs of what was a major change in plan for the contractors. There is an uncharacteristic 'bellying' of the bank of the canal on the right-hand side. Early plans of the tailrace show that it was to swing left at this point and enter the Shannon a few hundred yards upstream from where it enters today. The mouth of the tailrace would then have been positioned between Parteen-a-lax, alongside the Church of Ireland and the Cottage (then Crowe's house). As things turned out, due to many considerations, primarily cost, the river was routed down through the old school and Parteen House (Gloster's), passing by the front of the church and joining the Shannon between the Lax Weir and Captain Keane's Creek.

The new alignment, it was pointed out, would result 'in a saving of £45,000 in the partial development, but that this figure will be reduced to £17,000 in the

■ Excavator at work removing rock from the tailrace with rail ramp in background.

full development on account of the greater length of the tailrace to be widened'. This figure included estimates for expenditure problems associated with fissures in rock structure related to the rock foundation of the Shannon that might need special caulking. These geophysical problems were much less than those associated with option one and the German engineers had no hesitation in plumping for it. Neither had the Department of Finance, when it noted the saving involved.[1]

This major change of plan was seized on by the local wags, and commentators joked that the tailrace might even enter the Shannon at three different points including the little fishing village of Coonagh five miles away![2]

Had the original plan been adopted, the canal would have closely followed the course of the stream that came from Castlebank, flowed under a little bridge near the old school and entered the Shannon just above the little Protestant church. The stream now flows into the canal about two hundred yards up from the tailrace bridge on the left.

The first challenge faced by the contractors was to prevent this stream from flooding into the excavation works. They had to transport it over-ground in timber conduits or channels and return it to its original course, where it could safely run into the Shannon. Below the timber conduits, excavation of the bed of the new canal would take place; extreme caution was being exercised so that the conduits would not be damaged by the explosions and blasting of the rock.

The total length of the tailrace is approximately 1.5 miles. The contractors met with very heavy losses in excavations due to the quality of soil and to the massive boulders that had to be removed. This resulted in a considerable over-shoot of the budget, by £24,000.[3] It is easy to understand why they now wanted the least expensive route to the Shannon and to rethink the original plan. It is also clear why they wanted to renegotiate the original contract price with the government for any further excavation to be undertaken.[4] What they could not factor into the contract equation were the unforeseen difficulties they were to have with a number of people close by, whose lives had become a nightmare because of the blasting operations.

While there very well may be a comprehensive list detailing times and places of blasting, for the moment one has to rely on occasional references to it in reports of accidents and inquests, written complaints of residents along the length of the project and in cases of claims for compensation that ended up in the courts.

Work on the various sites went on around the clock, particularly when the contractors felt they were running behind time in the contract. Groups of workers kept locomotives going throughout the night, transporting waste and emptying wagons. The electricity supplied by the temporary power station enabled them to keep going at certain operations, including blasting. So much for a good night's sleep.

Apparently there were different kinds of blasts, depending on the work engaged in. There was the massive blast used to dislodge rock initially; anything up to 1000kg of amorite-gelignite could be used in one of these explosions. About six bore-holes, each eight metres deep, would be packed with two metres of explosive, two metres of sand with two further metres of explosive, to within two metres of ground level, for these blasts. There could be one of these a week – maybe two. These were terrifying occasions for people and animals alike. Apart from the noise, the ground would shake and rocks could land anywhere nearby. The combustion mix for these blasts was discontinued in April 1927, after an explosion in the tailrace broke windows in several nearby houses. One house was struck by a piece of rock weighing 14–16 pounds. It hit the outer wall with such force that a large piece of plaster was brought down from the bedroom ceiling! This happened at about one o'clock, when people were having their dinner. Rocks were also strewn about adjoining fields and on the nearby road.[5]

■ The Germans had their own experts in the use of explosives and trained some Irish assistants.

Once the massive blast was over, the excavators could move in and begin to shift the loose rock. After that would come lesser blasts, on a daily basis, in order to dislodge large boulders or to trim and align the canal sides and bottom. These, of course, would be accompanied with the familiar piercing siren warnings, given three times and each lasting half a minute, at intervals of a quarter of a minute. At night, in addition to the siren, a big red cross composed of red lights was switched on to warn locals of the impending explosion.[6]

At this remove, it is difficult to comprehend what life must have been like during those times. People would have been familiar with the beautiful, reassuring metallic sound of the mowing machine or the welcome dull gnawing rhythm of the threshing machine – sounds that would carry for a mile or more. But having to contend with the levels of constant dissonance generated by excavators, dredgers, rock borers and compressors, locomotives and their wagons, picks and shovels, and, of course, men shouting at each other, the traffic of forty big trucks travelling daily from Longpavement to the various sites, plus the disruptive effect of the explosions, must have made local people wish that they could just emigrate and forget it all, perhaps to return when it was all over.

Patrick Boyle, or Farmer Boyle as the contractors called him, was one of those. Technically, the family name is O'Boyle, but in all of the correspondence he is just referred to as Boyle. He lived with his wife and family on a line between old Ardnacrusha and McCormack's of Castlebank. The house was in the danger zone, just 50 metres from the canal. The McKies later occupied it while the Brownes bought the land.

■ There was a major detonatation of explosives weekly plus several lesser explosions for minor work.

As early as December 1926 Boyle's house had been badly damaged by the blasting that had begun during the summer. The walls had been cracked and boulders had damaged the roof. Prior to this he had been injured and had been awarded £5 compensation, which included damages for inconvenience caused by the contractors entering his land before the expiry of the government notice of three months.[7] On 13 December Boyle had called to J.K. Prendergast, the Resident Engineer, in the Strand Barracks demanding the protection for his house and outbuildings that had been promised some time earlier. The latter immediately contacted the contractors urging them to give Boyle the necessary protection, otherwise they would be held responsible in the event of a claim.[8]

By the middle of January the contractors had not provided the required protection so they were ordered by the Resident Engineer to cease blasting in the immediate vicinity of the house until they complied with instructions. By return of post they assured Prendergast that they would do so.[9]

During the following weeks, a timber roof was put on the house and blasting resumed. On the occasion of a big explosion at the end of April the contractors found themselves in a confrontational situation: 'Farmer Boyle was requested by our guard to vacate his house. He refused to do this. Dietrich went to Boyle and begged him to leave his house. After much discussion Mr. Dietrich was forced to threaten the vacating of the house by the police. Only then was the house vacated.' On 5 May 1927 at 4 pm, when blasting operations were again undertaken, Herr Schlapper had the same difficulties. As he could not come to an arrangement with the inhabitants, he addressed himself to the police in Ardnacrusha. The police refused to interfere, stating that the man could not be made to leave his private property. In the meantime the farmer had vacated his house, with the result that the blasting operations were carried out.[10]

Two days later the contractors were writing again as the blasting was getting very close to the Boyle's house. They were extremely concerned about possible delays and urged the Irish engineers to acquire the house immediately. 'As the house in question will shortly be in the danger zone of small blasting operations, which must also be carried out during the night, we would be forced to stop working …' They were of the opinion that the protecting roof that they had put up at the Resident Engineer's request would offer no security against accidents.[11]

Strand Barracks was concerned about the danger of serious injury to the family and considered building a shelter of wood strong enough to resist flying stones and boulders. For this it suggested that old railway sleepers would be got from the Great Southern Railways at a cost of two to four shillings each.[12] There was little realization then of the danger of creosote inhalation. At any rate, it was proposed that the Boyle family could sleep in this shelter at night. This was all fine in theory.

During the night of 24 May the Boyle property suffered more damage from the blasting. As a result, the contractors then arranged with Boyle that further protection should be provided for two of his outbuildings. When the contractors arrived Boyle would not allow them to work, charging that the protection was insufficient. The German engineers were of the opinion that the next big explosion would make the house uninhabitable. They would no longer guarantee the safety of the family and would not accept responsibility for blasting operations at night.[13]

The contractors continued to exert pressure on the Irish side that they should have the property vacated before work continued. This, they said, would be the normal way to proceed with contracts. Now the contractors found themselves faced with costly delays in their work and claimed accordingly. Mr Keehan, an engineer at the Strand Barracks, had the task of validating the claims. 'I return your contractors claim for time lost (blasting) with correction in red ink. The shovels have been delayed as far as I can see for the times stated. Mr Dietrich reported to me on the morning of July 20th, 1927 that Boyle refused to go out. He was milking cows at the time, but he went out after when he had finished.'[14] On a number of occasions in August he had refused to leave his house with the result that no blasting could take place. In one instance he assaulted a representative of the contractor, and in another he roundly abused Siemens' staff. On 17 August he refused to leave his house from 7.30 pm until 10.30 am so that blasting could not be undertaken. Writing to Professor Rishworth the contractors stated, 'We beg to inform you that this time Mr Boyle did not leave his house at all and, after giving Mr Boyle a warning in the presence of two witnesses, we had to start blasting without his consent. Mr Boyle said to our engineers that if any accident happened, the engineers employed on the tailrace building site would not go back to Limerick alive.' Boyle's behaviour was having a serious effect on the work of the contractors not only in terms of cost but also from a personal point of view: 'Yesterday evening Farmer Boyle refused to leave his house during blasting operations and this morning between 7.30 and 8.30 he again refused to leave his house. When our engineers tried to persuade him, he insulted them, and therefore they now refuse to risk a repetition of these occurrences especially as they are in no way obliged to do this.'[15] The contractors also included a bill for £900 for time lost as a result of Boyle's action. It was now up to the Irish engineers to do something about the situation.

Prendergast and his team at Strand Barracks were frantically searching around for a solution. They acknowledged that Boyle was owed compensation for the land already acquired from him and perhaps that payment could be made quickly by Finance. On top of that, a payment could be made which would enable the family to secure sleeping accommodation out of the danger zone.[16]

The place selected by Strand Barracks was a little cottage owned by Mrs Emily Crowe in land close to Boyle's. The latter refused to budge until the cottage was refurbished. He insisted that the bedroom floors and ceilings be boarded, that the kitchen floor be concreted, that all plaster be repaired and that the house be painted and whitewashed. Reading between the lines of correspondence, the cottage sounded like an extremely basic affair with earthen floors and rooms extending up to the thatch. Having been unoccupied for some time it had fallen into disrepair and was in no fit state for a family to move into. Local memory has it

■ Rock-borers taking a break from the back-breaking job of trimming the side of the tailrace.

that the building was not a dwelling house, as such, but a collection of outhouses where animals were kept. Boyle's conditions seemed extremely reasonable but the Resident Engineer thought that 'the above demands are excessive', particularly his request to have the cottage whitewashed! The fact that Boyle required £100 in disturbance money may have coloured Prendergast's opinion somewhat, but the rapidly amounting bills from the contractors because of delays due to Boyle's refusal to cooperate made his demand more attractive by the day. In spite of that, Prendergast refused to play ball and suggested to Rishworth that a sharp letter from the Department to Boyle might do the trick.[17]

In the background, Arthur Taylor, the government assessor, was working frenetically to bring about a breakthrough in the impasse. From correspondence and reports one can see that the machinery of bureaucracy could be ratcheted up when necessary. Taylor was about to travel to Limerick from Dublin and thought that if he could bring a cheque for compensation owed to Boyle the situation would be greatly relieved. Industry and Commerce quickly endeavoured to arrange this with Finance and thought it could be done. The cheque was released but the manoeuvre did not work, as Boyle would not cooperate, so the crux continued.

Shortly afterwards the 'strong letter' arrived from the Department of Industry and Commerce. In classic style it was addressed to Boyle as 'A chara', but there was very little 'cairdiúileachas' in the rest of it:

> You stated to Mr Taylor, this Department's representative, that you would go into Mrs Crowe's cottage if it was obtained for you. When it was obtained for you refused to occupy it. In the circumstances since you have seen fit to refuse to consider any of the proposals put before you the Minister wishes it to be made plain to you that blasting operations are necessary and cannot be discontinued: and further, that he is not prepared to make any additional offers to you. If you persist in refusing to leave your house during the actual blasting operations the Minister can accept no responsibility for any accident that may occur to you in consequence of the blasting. On the other hand the Minister is still willing to rent Mrs Crowe's cottage for your temporary accommodation. I am to add that the views expressed in this letter are definite and final.[18]

The letter seemed to have the desired effect and within two weeks it was reported that Boyle had agreed to move into Crowe's cottage for which the government would pay weekly rental of 30s. It was felt that Boyle would prefer to do up the cottage himself so he was to be paid £60 costs for that. He eventually moved into the little house with his family at the beginning of October but refused to sign an agreement with the government. As a result, he was not paid costs for refurbishment and this stalemate persisted right into 1928.[19]

In March 1928 Mrs Boyle contacted the Minister for Industry and Commerce complaining that their land was destroyed by stones scattered all over the place by the blasting operations. Having gone up through the system her letter came back down and landed on the desk of the Resident Engineer who dispatched one of his men to report on the matter.[20]

Two days later the report was ready. The surveying engineer said that he found Mr Boyle engaged in removing the stones with a horse and cart. Some of the stones were four stone in weight and some much heavier. There were several big heaps of stones about the land as the farmer had to clean up his fields repeatedly. The report also stated that there were eight cows in one field and that while all the outhouses had suffered badly, the principal cow-house was most severely damaged with over 50 per cent of the slates broken by flying stones.[21]

Eventually the Chief State Solicitor drew up a new agreement between the parties and in late June a payable order for £60 was made out to Mr Boyle. At least this would provide some consolation for the poor man as he toiled Sisyphean-

like through his land and made the cottage somewhat more comfortable for his wife and family.[22]

At the beginning of 1929 Shannon Power Development began to enquire as to when Boyle and his family could move back into their own house. The query centred on whether blasting had ceased in the neighbourhood of the house and what was the condition of the house itself. Apparently Boyle could not be instructed to return to his house until any damage done had been repaired, or alternatively, he had agreed to accept compensation and had signed an agreement discharging the Minister and the contractors from any further liability.

As it so happened, Boyle and his family had to stay in the cottage for some more time because further blasting was to take place in the canal adjacent to their house. On the west side of the canal there existed a rock ramp used by trains to enter and leave the tailrace at that point. These ramps were a feature all along the canal but were blasted out of it when the contractors had moved on to another section, and the particular ramp had become redundant. The side of the canal was then trimmed and the crusher processed all the rock. It was felt by engineers that the blasting of this particular ramp would further damage Boyle's house and there was no point in proceeding with repairs until that was over.[23]

It was March before the contractors had finished with that section and eventually the Boyles moved back into their own house. One wonders what their feelings were as they heard and read about the opening of the Shannon Scheme in July or what went through their minds as they stood outside their house and watched the first trickle of water flowing through the tailrace during the weeks of October 1929. No repairs had been done to their house, as a case was listed in Limerick's

Circuit Court on 18 October in which Patrick Boyle claimed £300 damages from the Siemens Bauunion. The contractors were very nervous about this case knowing that it was a very strong one against them and hurriedly contacted Shannon Power Development for advice as to whether they should settle beforehand. The case did come up for hearing but it was adjourned until the following February when it was further adjourned until late April. Obviously there were intense negotiations taking place behind the scenes and when the case came up for hearing on 17 July a note of settlement was registered in the Civil Bill Book.[24] Unfortunately, there is no detail as to what the settlement was but considering Boyle's record as a hard bargainer and the strength of the case, one could fairly presume that he got close to, if not all of his £300. He subsequently moved to another farm over at Quinpool/Ballycannon thus leaving behind him the trauma and anguish he had suffered for years.

Adjoining Boyle's farm was that of the McCormack family in Castlebank. Both families were neighbours up to the time of the Shannon Scheme but the tailrace split their lands and cut off whatever minor paths existed for walking from one place to the other. The McCormacks were somewhat farther removed from the blasting operations than the Boyles, but they did experience serious damage to their property from flying rocks, having one or two close escapes. Their story is as harrowing as that of the Boyles' and is detailed elsewhere in the chapter on land. They too had to vacate their home when blasting got very severe. Sometimes they refused to do so but most of the time they were amenable to moving.

Not so Patrick Clear. He ran the post office at Garraun, just above the present tailrace Bridge. He also ran a shop that sold confectionary and sweets and had some land. His story is an extraordinary one of move and counter-move with the contractors and government, of resolute determination not to be beaten even though he faced the full panoply of state machinery. As sub-postmaster for the area he was a government employee but this did not deter him from driving the toughest bargain that he could. Besides, considering that the post office was in the gift of the government, one of the relics of the old political spoils system, he was certainly jeopardising his chances of continuing as sub-postmaster by his action.

His house and shop and land were slap bang in the middle of the track of the tailrace and were to be acquired by the government. As early as 1926 the contractors were enquiring if notice had been served on Clear for his lands. Seemingly,

■ Patrick Clear at the door of his post office and home in Garraun, which was demolished to make way for the tailrace.

they had some difficulty with him in completing a temporary agreement enabling them to enter his land prior to the government acquisition of it. This problem arose because the contractors had already trespassed on his land without his consent, to make a borehole and to ferry material to the newly laid railway line; for this he had received £2 in compensation.[26]

Within a few weeks notice was served on Mr Clear for the land but not for the house, much to the annoyance of the contractors whose application had included the house also. The reason for including the house was that it was only 26 metres from the edge of the rock excavation and that, in their view, they did not think it possible to carry out the works without damaging the house and endangering the family. But for Siemens, speed was of the essence; otherwise the works would be held up at enormous cost. They had to have access to Clear's lands immediately and just could not wait for the statutory three months to elapse. In order to facilitate entry they made a private agreement with him to supply his house with electricity at a rate of 5s. per month in return for access. This was just before Christmas, 1926.[27]

Right through 1927 the difference of opinion between the contractors and the Irish engineers about the house continued. Rishworth, the Chief Engineer, was convinced that by exercising reasonable care, structural damage would be avoided and recommended covering the roof with timber and protecting the doors and windows with some form of lattice shuttering. He also proposed to the contractors that they should drill shorter holes for blasting and put in smaller charges than usual thereby creating smaller explosions and lessening any damage through vibration.[28]

The Germans were not convinced and they were right. In November, when blasting operations were being carried out close to Clear's house, cracks appeared in the gable wall which, as work continued, seemed to widen. Consequently, a wooden frame was fitted into the window in order to strengthen the wall. Then Clear told the contractors that the roof was no longer safe and that he was taking an action against them and the government. He also informed them that he now had to vacate his house even during the day.[29]

■ Clear's house covered with cladding to protect it from the blasting in the nearby tailrace.

The Department of Industry and Commerce was seriously concerned and Arthur Taylor became involved, recommending that the house be acquired immediately. Rishworth and other officials agreed with him. It was proposed that Taylor travel to Limerick taking with him the ordinary notice of acquisition for Clear's house. It was recommended though that acquisition should be by agreement if possible and that 50 per cent of the compensation should be paid in advance on condition that the family vacated the house immediately. Since it would take a week or two to put through the payment of 50 per cent, Taylor was authorized to offer a weekly sum of 30s. so that the family could accommodate themselves elsewhere at night. Seemingly the greatest danger was with the night blasting. The Department officials felt that when the 50 per cent was paid Clear could then move his stock-in-trade from the shop to other premises that he could provide out of the 50 per cent. They thought he had a plot of land outside the danger zone that would be suitable for that purpose.[30]

Again, all of the foregoing seemed fine in theory. It was only when Taylor arrived at Clears that he discovered that they did not own any other piece of land on which to build. The family were tenants of 'the chapel field' for over forty years, so Clear sought title to this piece of land. While awaiting the outcome of this request he occasionally opened his shop during non-blasting periods but ran a huge risk in doing so as the building was in such disrepair. In the circumstances, Taylor took it on himself to negotiate further with Mr Clear and eventually the latter agreed to forego the 30s. per week on the stipulation that he should be paid immediately the 50 per cent of compensation. 'I was able to secure these terms by using the lever of "the chapel field" which I understand the Land Commission has now agreed to assign to Clear.'[31] Again, matters looked rosy and settled. Clear was happy to give up possession at an early date and the contractors would finally be off the back of the Resident Engineer.

Four months later the crux persisted. Compensation had not been paid and Clear was insisting that the contractors continue to repair damages to his house caused by the blasting. Prendergast, the Resident Engineer, struck a curious note when he instructed the contractors not to do any more repairs on the house, as arrangements were being made to compensate Clear as he was quite aware that he was using the premises at his own entire risk.[32] This was no great consolation to the contractors who were then engaged in blasting rock for the foundation of the new bridge, bringing the centre of their activity even closer to Clear's house. The very real danger was that someone might be killed, resulting in a prolonged hold up of work.

Eventually, in early summer, Clear vacated his house but began to build a new one a short distance away on that portion of his land adjoining the old Barracks owned by Helen Ryan. This move confounded both the German and Irish engineers, as the spot he chose for his new house was in nearly as dangerous a position as his old one. Clear told them 'that the land is his and he has a right to do what he likes with it'.[33]

This plot was part of the larger parcel of land acquired by the government from him under the Shannon Electricity Act, 1925, but handed back to him in November.[34] Professor Rishworth wrote a craftily worded letter to Clear:

> ... I warn you that it is dangerous to erect any structure on the proposed site until the blasting operations being carried out by the contractors of the Shannon Scheme have been completed. The recent compensation paid to you includes consequential damages attributable to blasting and no further claims for compensation will be considered. I am to add that no objection will be raised by this Department to your using 'the chapel field' for the purpose of building a house for yourself thereon.[35]

Fortunately for all concerned, and particularly the Clear family, there was no one present in the new building when nearby blasting sent a stone through it on 8 June, causing a lot of damage. Michael Tynan, solicitor for Clear, wanted to know from the contractors whether they were prepared to repair the damage and what protective measures they were going to employ in the future. The contractors were furious, but were instructed by the Resident Engineer not to undertake any repairs to the building and not to reply to Tynan. This was purely a holding measure as there was the possibility that Clear would sue the contractors. In the event he did not.

At this point the relevant file runs out, apart from one final letter from the Resident Engineer to Professor Rishworth about the acquisition of land. It was dated 15 October 1929. The contents of the letter indicated that Clear was being offered an option on a number of plots in the area of the lower tailrace but none of them proved viable, mainly because of difficulty of access.[36]

In the end a new shop was built for him alongside the chapel, which he ran well into the forties. He and his family lived for a time in the house he had constructed by the old barracks before moving to a more spacious and comfortable home just below the chapel. Thus ended their extraordinary sentence of insecurity, deprivation, loss, harassment, and even terror. The Scheme meant that they lost not only their ancestral home, but that their livelihood was sorely restricted by the temporary loss of the shop and the permanent loss of the post office. The only remnant of their house that remained, right up to the late 1950s, was a little collection of raspberry bushes on the bank of the canal from which local children collected the fruit.

The opening of a new post office at the workers' camps at Ballykeelaun in September 1926 meant that Clear would probably lose this facility anyway, as the greater volume of business was at the camps. The pay office at Parteen, near the lower tailrace, was also closed at this time. Presumably the transfer of Ardnacrusha post office from Garraun to the camps, a mile north of the village of Ardnacrusha, dates the formal change of the name Ardnacrusha to Parteen. Thomas Ryan, formerly clerk of the Limerick Board of Guardians, ran the new post office.[37]

The Shannon Scheme was an enormous leap forward for the country in so many ways. It represented a remarkable triumph for the government, the contractors and the country as a whole. Everyone was a winner and wanted to be part of the celebrations. All the more reason then to mark the memory of those who unfortunately were casualties of progress but whose dignity, bravery and resilience earned them a significant place in a great story.

7 | Counting the Real Cost

It was only inevitable that some lives would be lost during the Shannon Scheme. It was a succession of building sites, miles long, with thousands of workers, ant-like, beavering away against the clock for the completion of the project.

At one end of the canal, thousands of tonnes of rock were being blasted into the air by controlled explosion, while at the other, millions of tonnes of clay were being gouged out of the ground by enormous machinery, the like of which was never seen before in this country. 'Health and Safety' was less than basic and the lack of labouring know-how was costly. During the three and a half years of construction thirty-three men lost their lives working on the various sites, while another fourteen people died in related accidents, not counting the nineteen crew of the *SS Arabia* lost in a storm in March 1925 en route from Germany to Limerick Docks loaded with rails, trains and other heavy machinery. A number of other workers died of natural causes and one was murdered. The total amounts to a colossal fifty-two deaths, a horrifying figure for us today and one that even caused concern at the time. But in the absence of comparative figures for similar undertakings it is difficult for us to judge how bad it really was.

In March 1926 the first major accident occurred on the site. A labourer was killed and three others were injured, two seriously, while ballasting track on the Blackwater-Ardnacrusha road. The railway gang lifted the track by means of long

■ A labourer tied by a rope engages in the dangerous work of trimming the side of the tailrace.

wooden levers that took purchase on the track under the rails. As the train approached, the German foreman ordered the removal of the levers. Unfortunately, one lever was only partly withdrawn and the train in passing, caught it, flinging it forward with great force. The lever struck one workman tossing him onto a track eight feet away. Then it got stuck in another part of the train and hit two more of the gang. All three were hospitalized, but one died later. He was William Byrne of 3 Pearse Terrace, Arklow, and the first in a long line of men to die on the Scheme.[1]

An inquest followed acquitting the contractors of negligence. However, of more interest is the internal report of the Resident Engineer, J.K. Prendergast. In it are the first recorded statements of German dissatisfaction with the quality of Irish labour, which in fact was a constant of theirs throughout the entire period of the Scheme. Chief Site Engineer Stolberg did not mince his words:

> From a purely legal point of view, it is hardly possible to make any person responsible for the accident; the actual cause simply is, that the man (the Irish train driver) had no experience in the work at hand … It is beyond doubt that our Irish workmen

will in time acquire sufficient experience in earth works to prevent similar happenings in future ... It can be safely asserted that the accident would not have happened if the locomotive had been driven by a German, since engine drivers with experience in civil engineering work exactly know where their engine can pass without obstruction. The Irish locomotive drivers may have driven locomotives for a long period and without reproach on the railways of the great companies; but they still have to learn to work railways on a building site, and such experience cannot be acquired in a short time's practice. They must be prepared for numerous causes leading to accidents of a purely technical nature, or, what is worse, accidents endangering the lives of their fellow workers ... to safeguard against further accidents, I should urgently request you to concentrate your efforts on ensuring that the staff of German foremen and machinery personnel is by no means reduced, but rather further increased. There remains no doubt that it will still require some time, (how long cannot be estimated at present), until Irish workers are sufficiently trained before a reduction of the German personnel can be considered.[2]

With Stolberg's report went a covering note from his boss Heintze:

... we wish to stress again that we are forced to work with unskilled workers from the country who have never worked in a big civil engineering enterprise. This type of labour requires constant supervision by a sufficient staff of foremen and skilled workers who have had training and expertise on large working sites.

Undoubtedly the Irish labourers had little or no experience of building sites and of the skills involved. But on the other hand there are stories of men getting jobs as locomotive drivers, the only qualification necessary that of being able to drive an engine up out of a big pit close to O'Brien's Bridge, where the fishing lake now stands. Reportedly, lads as young as sixteen got jobs as train drivers and one of them, from Castleconnell, was involved in an accident at Blackwater. However, with nearly sixty trains and a thousand wagons on their books, all imported from Germany, it is not too difficult to see how safety corners had to be cut in order to keep the show on the road, particularly as there were enormous penalties if the contract over-ran.[3]

It is hard to comprehend that the work needed a thousand wagons. It is only when one takes into account the statistics of the undertaking that one gets an inkling of the enormity of the challenge. The site took in an area of nearly nine miles long involving the excavation of 1,350,000 cubic yards of rock and the placing of 9,500,000 cubic yards of canal excavation in embankments. The wonder is that it was all done with a thousand wagons!

■ Working with the trains proved to be the most dangerous of all occupations.

Sadly, however, these wagons exacted a high price in terms of human life. In order to be emptied they had to be tipped on their side and held in that position until the operation was completed. Frequently they swung back unexpectedly, jamming and crushing men between the trap and the floor of the wagon.

The first fatal accident of this kind happened at Parteen on 22 April 1926.[4] There were nine wagons attached to the engine. The protocol of emptying was that the foreman would order the brakeman to insert the poles that would prevent the wagon from tipping back again. Men would 'stand clear' while this was being done. Once completed, the wagon could then be emptied safely.

On this occasion, according to the Irish foreman, it appeared that the wagon was not secured by a pole. He shouted to the men to 'stand clear' and went to find a suitable pole. On his return he accidentally touched the tipping lever with the pole causing the wagon to swing back crushing the labourer, Mark McDermott from Leitrim, who had approached the wagon at the last minute. Other labourers were of the opinion that he had pulled out the pole for use elsewhere and that the wagon tipped back of its own accord, catching his colleague and killing him.[5]

By the end of 1926 there had been nine fatalities on site, five of these train-related. The contractors were really concerned about this, even alarmed, and they begged the Irish side to do something about it. They contacted Professor Rishworth at his head-office in Merrion Street, Dublin, 'This figure is very large even for

■ Professor F.S. Rishworth of Galway University, who became Chief Civil Engineer with Shannon Power development.

an undertaking on so large a scale and with so many machines as the Shannon Scheme. As contractors we shoulder a very great responsibility in regard to deaths and other accidents and therefore wish to put our views on the matter before you.'[6] They pointed out that they had observed all the strict German regulations for building sites and had also observed, what they considered, the less strict Irish regulations. Even when the Irish had insisted on additional precautions, they had implemented those as well. They pointed out that they had continually brought to the workmen's attention the dangers of the building site through posters, notices in the Pay Offices as well as constant verbal reminders:

> The numerous accidents amongst Irish workmen can only be explained by the fact that they are inexperienced in work on large building sites …. We are convinced that we could diminish accidents if we had a larger German staff. We notice the lack of experienced personnel on our machines which suffer considerably more than they do on our other building sites. Our German personnel is strained to such an extent through having to work overtime and to watch over the whole work that we can attribute the many accidents and the considerable cost in repairs to this cause. We therefore urgently beg you to approach the government in this matter and ask for the sanction of a larger number of Germans. The demands that fewer Germans should be employed, which are continuingly being put forward by the Trade Unions and other political organizations, put a heavy responsibility on the latter for the many accidents which continue to occur.[7]

Paddy McGilligan and the government had to be careful on this one as allegations were being tossed around about Germans doing work that could validly be done by Irish craftsmen and that German labourers were being employed on sites. There may have been a basis for the first allegation as many of the German workmen were carpenters, fitters, mechanics, smiths etc., but the question has to be asked as to the volume of Irish craftsmen available, particularly those with relevant experience. On the second allegation that the contractors were employing German labourers, evidence would indicate otherwise. The 800-odd German

■ Many men were killed and others seriously injured while tipping and emptying wagons.

workmen (excluding engineers) are all listed as having a particular skill or trade whether it be explosive expert, electrical fitter, dredger driver, belt saddler, rope-maker, concrete specialist or whatever.[8]

Apart from the labour question, Paddy McGilligan was coming under pressure in the Dáil about the number of accidents, particularly fatalities. In his own feisty, combative style he frankly presented the statistics regretting the catalogue of deaths but keeping his powder dry on German demands for increased numbers of staff.

By the end of 1926 there were approximately ninety German foremen on site. Over two dozen more were brought in early in 1927, along with nearly twenty extra train drivers and many other tradesmen. However, while the number of fatalities was fewer during the year (three fatalities – two train-related) the rate of accidents, many of them horrendous, was considerable. Heintze, Siemens' Chief Engineer, was still corresponding with Rishworth's office about it and specifically about Irish disregard for their own safety or for German instructions:

> We have repeatedly warned the workmen verbally and in writing that they should not get into the wagons for the purpose of emptying them. Special shovels with long shafts have been provided for the purpose. We have also supplied mechanical appliances which are a safeguard against the wagons falling back. We cannot arrange to enforce our instructions forbidding the Irish workmen to climb into the wagons without continually keeping watch on the men and continually repeating our instructions .... We do not consider that the language difficulties are an obstacle preventing German foremen giving a timely warning .... In cases where a simple shout can prevent an accident, it does not of course matter in what language the foreman shouts.[9]

In support of his general argument about Irish lack of experience being a contributing factor in accidents, Heintze quoted the Ocean Insurance Company as having pointed out that accidents involving German workers were much fewer, relatively speaking.

The final point in his letter is interesting in that it indicates how Rishworth was putting pressure on the contractors to employ more Irish personnel. Heintze mentioned how Irish staff was doing the checking of hours, wages etc., and that Siemens had even employed more locals for clerical work than they required. He did admit though, that their accounting system was German and therefore could only be operated by German clerks. He then suggested that Rishworth would be most welcome to send an expert to their offices to check out any matter of concern.

German angst about the Irish handling of the wagons erupted again at the end of 1928 when three men were crushed into the ground as a wagon loaded with earth swung back suddenly, striking them and dumping its contents on top of them. The men were on night shift on the Newtown-Clonlara section of the Scheme when the accident happened. The men had to be dug out from under the wagon but were pronounced dead at the scene by Dr Myles McSwiney, the camp doctor.[10]

That accident was one of the worst of the Scheme and was quickly followed by another horrific 'tipping' mishap when the railway sleeper being used to keep the wagon tilted, snapped and broke, crushing the young man who was involved in the emptying operation.

The contractors again issued instructions on safety measures to every foreman and ganger on the site along with illustrations as to how the job was to be done. Moreover, all supervisors had to sign a declaration form acknowledging receipt of these instructions. But there was more to the story.

The Irish Chief Engineer, Professor Rishworth, was outraged. He claimed that the German measures were inadequate and in breach of a recent agreement. He sent a rocket of a letter to Siemens Schuckertwerke:

> … you agreed on the occasion of a previous fatal accident to provide an automatic locking arrangement which would come into operation at any angle and also provide steel props. Neither of these safeguards would appear to have been in use on the truck that caused the fatal accident. The Minister takes a very serious view of your failure to carry out the undertaking made with this Department … I have to direct you to issue notices to the building sites which should be posted up at the Pay Office warning men that under no circumstances must they enter the truck or put themselves in danger while emptying partially tipped trucks, and at the same time warn your foremen and gangers that even if it should cause delay, the above order must be rigorously enforced.[11]

■ Train derailed at Clonlara.

The year 1928 was an appalling year for train accidents and it closed with an horrific death at Parteen on 27 December. Daniel O'Brien from Templenoe in Co. Tipperary, who had served with the Flying Corps in the Great War, was returning to work after dinner at about 2 pm. The thirty-five-year-old rock borer went down the ramp to the floor of the tailrace, where two trains were servicing the steam shovel, removing rock that had been blasted earlier. He left the pump house at the side of the ramp, crossing the railway tracks going up out of the canal and made his way to the second railway line right in the middle of the floor of the site. He walked along the tracks for about thirty yards and then apparently stopped to watch a loaded train make its way up the ramp. What he did not realize was that there was a second train with three empty wagons coming along behind him on the line he was standing on. It was making for the steam shovel up ahead. The engine was at the back, pushing the wagons along. The driver had seen O'Brien leaving the ramp and had whistled to him to keep clear. But there was little chance of anybody hearing a human whistle with the ambient din of the two steam trains, a steam shovel and dozens of men with picks and shovels and boring machines. At any rate, the driver of the second train was unsighted by the wagons in front of him. The first one hit O'Brien killing him; his body was dragged thirty yards farther up the track, mangling it badly.[12]

Johnny Cusack of old Ardnacrusha village (Parteen), now the sole Irish survivor who worked on the Scheme, remembers the incident well. He was a young messenger boy on the site and ran up the fields to get Fr Kelly, the recently appointed curate in the parish, to attend the body that had been torn into three parts. This accident happened at Section 50 in the tailrace cutting situated between Boyle's house and Davidson's Bar.

While work was in operation, many commentators used to marvel at the sight of the Clare countryside at night, brightly lit, a foretaste of what was to come when the Power Station was up and running. During construction, the Temporary Power Station provided electricity. Cables ran the length of the head- and tailraces, not just providing light but powering machinery and the extensive telephone system. Even some of the rail system and the locomotives were powered by electricity.

Ironically, a young German fitter was the first to be electrocuted in July 1926. Leonhard Gernsbeck was working in Clonlara where he removed a switchbox and came in contact with a live wire. His wife, Babette, came from Germany for the funeral and he was buried in Doonass.[13]

■ Dr Myles McSwiney, the camp doctor for Siemens.

In November of that year, twenty-eight-year-old Thomas Whiteside, from Scotstown, Co. Monaghan, was the first Irish worker to be killed by electricity. He was an attendant at the hydraulic pumping station in Parteen and was found dead on a Sunday morning just alongside the pump-house. He had come into contact with a high-tension wire that had snapped and fallen to the ground during the night, which had been very stormy. Whiteside appeared to have been a bit of a character. Reportedly, while he was out on a spree on a Saturday night in Limerick, shortly before his death, he posed as a plain-clothes detective and made the driver of a car get out and clean his number plate. He then warned the driver that he was going to hand him over to the next uniformed Garda that he met! Unfortunately for him, the next Garda he bumped into was an inspector who recognized that Whiteside was not a colleague. Poor Whiteside himself was arrested and was out on bail when the accident happened. He left a young widow and four children in Monaghan.[14]

An extraordinary occurrence caused the death of a sixteen-year-old messenger boy at Clonlara in August 1928. James Markham of Thomondgate and his young friend Matthew Kelly were fooling around with a bulb socket on a lamp. They were putting their fingers into the live socket experiencing the resultant shock. A German supervisor warned them of the possible danger and told them to desist. The boys ignored him. They then tied a wire to the socket and attached it to a vice outside the window of the office. They excitedly continued with their experiment not noticing that the wire outside the window was getting wet from a shower of rain. Suddenly young James could not free his finger from the wire until somebody turned off the lamp.[15] He collapsed on the floor of the office, dead. These details emerged from the subsequent inquest, but folklore has it that the boys had tied the wire from the socket of the lamp to the knob of the office door and that the intended victim was the next German to open the door. At any rate, of more interest still was the session in the Limerick Circuit Court when his father Stephen made a claim under the Workmen's Compensation Act for £300 for the loss of his son who earned 30s. per week as an employee of Siemens Bauunion. Curiously, Arthur Cox, one of the biggest and most brilliant legal guns in the country at the time and advisor to Siemens, was in court. His presence was unusual in itself but it was indicative of how seriously the contractors were taking this case. It emerged in evidence that it was young Markham's friend, Matthew Kelly who had tied the wire to the vice. Counsel for Siemens argued that Markham's death did not arise out of his work as a telephone operator. Unfortunately for the family, Judge McElligott had no hesitation in dismissing the action![15]

The Markham family certainly must have felt aggrieved with the result of their claim, particularly in light of an earlier claim by the family of Bridget Walshe, a typist with Siemens Bauunion who stayed with her aunt, Mrs Kearney, in Parteen and used to cycle to Clonlara where she worked. She was knocked down and killed by a Siemens vehicle at O'Connors' Cross in May 1926 on her way from work.[17] In February 1928 her father Robert, from Ballyhakill, Co. Limerick, brought his case to court claiming £300 compensation from Siemens Bauunion for her death. The family were awarded £250.[18]

Considering the novelty of electricity for so many of the workers it is remarkable so few fatal accidents were attributable to it. Up to the opening of the Scheme in July 1929 there were only three deaths in all. There should have been a fourth, but inexplicably, the man lived. This accident happened at Blackwater when a labourer was going down the headrace bank and got bogged down in some soft

■ Screeding the sides of the headrace.

muck. In order to get out he reached up and grabbed one of the cables supplying electricity to big machines at 3000 volts (3-phase). He immediately began to scream. One of the German electricians, on seeing the man's plight, had the presence of mind to throw a fish plate across the cable held by the labourer and another cable so that the lines shorted, rendering them dead. Onlookers held their breath, but the lucky man survived. He pulled himself out of the muck and, apart from some slight burns to his hands where he held the wire, he was none the worse for wear, resuming work immediately. The fact that the man was in no way insulated through the wearing of rubber boots or gloves, and that he was half buried in muck, make this episode seem a rather extraordinary case of immunity from electric shock.[19]

Not so lucky, though, was a young Irish electrician named Patterson, who was badly burned at O'Brien's Bridge in June 1927. He was working on the high-tension wires when a German electrician named Winmer unwittingly turned on the current. As a result Patterson was severely shocked and was rushed to hospital, where an arm and both feet were amputated. Apparently, further surgery was needed and his colleagues were pushing for his transfer to a Dublin hospital where treatment would be better.[20] Unfortunately, there is no record in the relevant file of the outcome.

The Shannon Board of Control, set up by the government to monitor progress and iron out any problems that might arise during the construction period, was particularly concerned about accidents and compensation. In 1926 its secretary, James Fay, contacted the Resident Engineer's office looking for details of fatal accidents, accidents by which the worker was incapacitated for further work, and compensation. Prendergast's reply is interesting because of the information it gives on the procedures involved and the attitude of those administering them:

> As regards this query, there is considerable difficulty in determining cases in which the worker has been totally incapacitated for resuming his ordinary work … I have heard of only one case in which the claim for total disablement has been made and I have not been able to ascertain if this claim has been allowed. Naturally the Insurance Company, Ocean Insurance, and the Contractors must be very careful to avoid allowing such claims until they are fully satisfied that total disablement has, in fact, occurred. You will realize that a number of accidents have occurred in which the workman has not been able to resume his duty, but it is not possible in a short period to determine whether the non-resumption of work is due to dis-ablement as a result of an accident.[21]

With regard to compensation, the Resident Engineer replied that

> … as required by the Contract, all workers are insured and that the Insurance Company has given compensation in all cases where claims have been justified and in accordance with the 'Workmen's Compensation Act'. The Insurance Company are naturally reluctant to give their detailed figures of compensation, should there be any danger of same being made public, but their representative here informs me that at any time their Manager in Dublin will be pleased to discuss this matter with the Minister or his Representative and to give him all figures confidentially…
>
> I understand further that in the case of fatal accidents, all claims for insurance have been made by the Insurance Company. In the case of other accidents the Insurance Company have only found it necessary to oppose a few claims, and where they have done so, the Courts have upheld the Insurance Company's case.[22]

In fact, the Limerick courts had a busy time during the Shannon Scheme, with a variety of case material going through the system. Usually compensation cases were processed through the Circuit Court where claims were heard under the Workmen's Compensation Act. Judgments were handed down fairly quickly and

■ Massive landslide at Clonlara in 1927, one of many.

while there may have been justice, there was little mercy. For example, Michael McKelvey, a foreman on the Scheme who earned £5 10s. a week, was injured and incapacitated. He was a carpenter by trade but could not function as such any more. When his case came to court the judge awarded him 30s. per week, but added the warning rider, 'not for all time'.[23]

It was only in such an occasional comment from a judge that one could begin to surmise what the 'official mind' was like. Undoubtedly the government was strapped for cash and could not afford generous payments to injured parties. But were strict riding instructions given to judges in Workmen's Compensation Act cases? Hard to tell. However, it is obvious that they borrowed pace from the government line and reflected ministers' attitudes in their comments. For example, a labourer, Denis Hishon from Charleville, appeared before Judge McElligott in Limerick in November 1929, claiming compensation for injuries he had received to his hand while he was working on a stone crusher. The accident had necessitated surgery and he was laid up for five weeks. The judge dismissed the case and he told Hishon that he was one of those who thought he could get money for nothing!

On the same day there was another interesting case, that of a German, Herman Gunther. He had been injured in August 1928, when a pile of debris fell on his feet, incapacitating him completely for a time. He returned to Germany and had surgery. He came back to Ireland in February and resumed work in April.

In court he was seeking compensation for loss of earnings from February until April. Judge McElligott refused his application with costs![24]

However, he was not entirely a curmudgeon on the day, conceding 35s. per week for two weeks to Dominic Barron who had been struck by a bogey and was out of work for five weeks … Even the justice element was harsh.

With 1,350,000 cubic yards of rock to shift and 1000 tonnes of explosives with which to do it, one of the most awesome, and at times frightening, experiences of the Shannon Scheme was the blasting. People were killed, others seriously injured. Animals were terrified – horses ploughing went into a frenzy and cows 'threw' their calves prematurely. Lumps of rock were sent flying into the air and some pieces landed over two hundred metres from the site of the explosion. Nearly all houses were seriously damaged along the length of the head- and tail-races even though many of them had protective timbers covering the roof and walls for the duration of the blasting. There is even a record of a lump of flying rock weighing about seven pounds travelling over two hundred yards and crashing through the roof of a house in Parteen, landing in the kitchen where the lady of the house was sitting! Fortunately, she was not injured.[25]

Considering the size of the operation, the number of listed German explosive specialists was relatively few. Franz Reinschmidt was the first of these to arrive in December 1925 and five or six others followed him. They were accompanied by a number of rock drillers whose foreman sported the splendid and illustrious name of Fritz Luther.[26] He must have caused some curiosity among the local monolithically Catholic population of the time!

While the blasting caused enormous physical and social disruption, it was an operation that was well regimented, generally speaking, and executed according to a very specific drill. There was an elaborate system of warning hooters and lights that indicated that a blast was imminent.[27]

The first recorded incident was in March 1927 when Michael Dooley, at Parteen, suffered serious head injuries when he was hit by a piece of flying rock after the explosives went off. There were a number of other close shaves for people including Garda Jones of Clonlara who, along with other spectators, was watching the blasting there. He was caught by a piece of flying timber, but his injuries were not that serious.[28]

The most tragic event associated with the blasting was the accidental death of a German, Josef Horner, in October 1928. It was a bizarre incident and happened as the tailrace was being excavated near Parteen. The accident underscored the importance of following protocol which, on this occasion was inexplicably set aside. Horner was an explosives foreman. He had trained an Irish assistant, Eugene McSweeney. As Horner was down in the bed of the canal preparing the wires for the charge, McSweeney, up on the bank, threw the switches activating the detonators. The unfortunate Horner was close to the centre of the explosion and caught the full force of it. He was thrown high in the air and killed instantly. No warning hooters had been sounded.[29]

■ View of Blackwater Bridge and giant crane, which fell trapping Louis Krieger and Anton Grossman beneath.

At the inquest McSweeney, of Cecil Street, Limerick, could not explain why he had interfered with the switches, as it was not his job to do so. He was subsequently charged with manslaughter but was eventually acquitted.[30]

As work on the various sites neared completion, some horrific accidents occured that involved the dismantling of the heavy machinery being shipped back to Germany. Two young men, John Larkin from Galway and Joe Twoomey from Kilmallock, were taking down a giant dredger at O'Brien's Bridge when part of it fell on both of them, killing them instantly.[31]

■ Anton Grossman.

The catalogue of accidents is practically never-ending but one has to wonder about those who suffered serious injury and the type of care, if any, that they received subsequently. There were those who were pulled into machines and were seriously mangled; there were those who lost limbs; others were incapacitated through massive spinal damage and horrific internal injuries.

Michael Dooley (already mentioned), whose skull was fractured by the flying rock, was hospitalized and was not expected to live. James Cunneen, another labourer, who had been caught in a drilling machine, was moved to the City Home and his condition pronounced precarious, but one never heard of him or Dooley again. And then there was Patrick O'Reilly from New Ross who, while servicing a dredger, had the flesh ripped from his hip and his rectum torn away. How did he fare out afterwards? Considering the rather undeveloped state of medicine at the time, one can only speculate and fear for the worst. This is no reflection on the heroic camp doctor, Myles McSwiney, and hospital aides at Ardnacrusha, who did the best that could possibly be done for victims. But, like today, hospitals cost money. According to the Shannon Scheme contract, Siemens had to pay £2 2s. to Limerick County Borough Hospital per patient per week; the contractors strongly objected to paying for any more than a two-week stay. They were not alone in that, as the local authorities were 'financially challenged,' as it were, and just could not afford to keep people in hospital. Presumably people were discharged as quickly as possible and they just returned home to live out the remainder of their lives as best they could, living on whatever little they got by way of compensation under the Workmen's Compensation Act or from the Ocean Insurance Company.

And it was not just the Irish workers who experienced such hardship. The Germans also had a pretty tough time. The case of Herman Gunther is just one. A much more poignant example is that of Anton Grossman, an electrical fitter, who was seriously injured at Blackwater Bridge in March 1928. He had gone to the assistance of his colleague, Linus Krieger, when a giant crane collapsed on them both, trapping them underneath. They were rushed to St John's Hospital, but it was found that Krieger was not as seriously injured as Grossman, whose chest had been crushed and he was bleeding from the ears indicating a skull fracture. As a result of the accident he lost the power in his right arm, his torso bore many scars from the deep wounds, but of more consequence, his brain was occasionally affected by the cerebral contusion of the accident. Like Gunther, he too returned to Germany for surgery, and returned to Ireland shortly afterwards. In February 1929 he claimed compensation under the Workmen's Compensation Act and the decision entered in the register was 'Agreement Signed'. There was no detail as to what the agreement was, but his son recalls that his father received £500 in settlement, much of which went on medical expenses afterwards. When the Scheme finished he did not return to Germany, having married Bridget Collins from Labasheeda in Clare. He worked for a time with the ESB and with Knocklong Creamery, before eventually succumbing to the long-term effects of his injuries that had grievously affected him, causing him to be hospitalized for protracted spells. He was in his fifties when he died, leaving a widow and three children.

Appendix 2 contains the names of those who died or were injured on the Scheme and a comprehensive listing of the awards made by the courts under the Workmen's Compensation Act to those who suffered injury or to a deceased's next of kin. The story of the Shannon Scheme and of those who lived and worked for it has been versioned in many ways and frequently told down the years, but only now are we getting around to remembering those who died for it.

8 | Keeping the Lid on Things

President William T. Cosgrave performed the opening ceremony of the Shannon Scheme on Monday, 22 July 1929, at Parteen Villa, O'Brien's Bridge. It was a very wet day, but that did not dampen the excitement about what had been achieved or what was to follow. During the next few months water was siphoned from the Shannon and slowly passed through the sluice gates into the headrace in preparation for the generation of electricity at the power station.[1] By the end of August work on the construction of the tailrace was completed.[2] In September the giant turbines were tested.[3] On 21 October those areas of the country that were adequately wired to receive electricity did so.[4] Gradual scaling down of the construction works had been under way for some time and employment numbers quickly dropped from 5000 to 2000 men.[5] With the reduction in these figures went a commensurate lowering of the crime rate associated with the Scheme that, at one time, almost dominated the business of Killaloe, Limerick and Ennis courts. Bertrand Russell may have been correct when he wrote that 'criminals, in the legal sense, seldom have much influence on the course of history,' but there is little doubt that records of the circumstances of their crimes as documented by the courts tell us a lot about their world and the social conditions in which they lived and worked.

While there were a few celebrated cases associated with the Scheme, most of the crime was of a petty nature, frequently associated with drink. There were

■ The superstructure takes shape with one turbine nearly fitted.

twenty-six pubs within a two-mile radius of the works, not counting the clubs in the camps and the various shebeens that did a roaring trade in poitín.

One of the earliest reported cases concerned a group of eighteen German engineers, scientists, electricians, servants, and, of course, Dr Myles Mac Swiney, the camp doctor, living in Doonass House. The majority of the Germans working on the various building sites lived in nearby wooden houses and huts. They had their own club. They also had their own school for their children, run by Herr Stumer, and their religious needs were looked after by Rev. Jupp from Roscrea.[6] Those in Doonass House were perceived as the aristocratic grandees of the German contractors and had rather exclusive arrangements for themselves, so much so, that Doonass House was looked on as the epitome of up-market hauteur. It is difficult to imagine what must have been for them a heady cocktail of genuine culture shock and outrage when they received a summons under the Shebeening Act from the local gardaí on 26 June 1926, as a result of which their fine cellar of wine, totalling over 1500 bottles and valued at £700, was seized.

A month later the case came up for hearing at Limerick County District Court. Baron Gravenitz, a senior executive of Siemens Schuckert, gave evidence. He stated that all of the drink belonged to the company, which had taken over

■ A bus could fit through one of these turbine intakes.

Doonass House for its use. He told the court that in fact there were two cellars, one for residents and one for visitors, who were mostly German callers. Acting District Justice Conroy did not have much mercy on the Germans as he ordered that all the drink be confiscated. He imposed a fine of £10 on Ernest Fiege, the mess steward, for selling liquor without a licence on 1 March and on the day of the raid.[7]

Another case that caused a certain amount of amusement and revelry among the navvies on the works concerned a consignment of Murphy's stout that went astray. The consignment, meant for Anselm Taylor, a caterer in one of the canteens at Ardnacrusha and subsequently the owner of a pub at O'Brien's Bridge, disappeared from Longpavement Station, having been collected in a lorry by someone other than Taylor. The latter then brought an action against Great Southern Railways and was awarded £25 compensation by Judge Mc Elligott in Limerick Circuit Court. Reportedly, it was one of the 'wettest' summers ever in the area.[8]

While these cases may have brought a certain amount of hilarity into workers' lives, the increased rate of traffic in the courts did not amuse the judges. The increase in crime was putting a lot of pressure on the system. At Killaloe District Sessions, Judge Gleeson declared that he was determined to put an end to the scandal of drunkenness at O'Brien's Bridge, which had become notorious, there being more cases there than in the whole of Clare. He accused the Shannon Scheme workers of being mostly responsible.[9]

Curiously enough, it was predominantly the Germans who were making the headlines in this regard. For example, during the lead-in to Christmas 1927, two of their mechanics tried to get into Sadlier's pub in Catherine St, Limerick, just after midnight. They were refused entry, but being already somewhat 'tired and emotional' after a serious crawl, they began to create trouble on the street outside. Inevitably, the guards were called. On arrival the Germans assaulted them and one poor guard had to face Christmas without his two front teeth.[10] The season of goodwill was still in full swing when another group of Germans left a pub in the city late at night. 'Feeling no pain', one of them decided to entertain the locals with a rendition of The Watch on the Rhine. One of his countrymen took grave exception to this and a vicious fracas ensued. Again, the posse of guards duly arrived and after firing a few shots they managed to break up the row and arrest five Germans – but not before one of their number, a detective, sustained a bad gash over the eye.[11]

While the Shannon Scheme was looked on by many at the time as a 'Free Stater' project, making the government extremely sensitive about security matters, there was little direct threat to the project itself. Even when the enormous stores at Ardnacrusha went up in flames in September 1927, and subsequently the same thing happened to the massive storage depot at the railhead in Longpavement in May 1928, there was no major stepping up of security as a result. The paranoia of the authorities never really translated itself into a commensurate show of force on the ground, even though it was discussed ad nauseam in certain quarters.

The larceny of food, clothes, bicycles and tools was by far the most common crime on the Scheme, but occasionally it took on more serious proportions. For example, in November 1928, John Hogan, John O'Neill and Ed Toomey appeared in Ennis District Court for attempting to rob £2000 from the pay office at Ardnataggle, O'Brien's Bridge. The robbery had been planned in the disused Blackwater Mill, but the men were captured before they had completed the job.[12] Another hold-up that ended in failure was that conducted by Terence Connolly and James Kelly, employees on the Scheme. Armed and masked, they held up the mail motorcar at Kilmore, Broadford, in November 1927, robbing

■ Two fires destroyed the massive stores at Ardnacrusha and Longpavement.

£100 and the car that they later abandoned at Hassett's Cross, near Killeely. Subsequently they were captured and brought to trial.[13]

Some appearances in court were the direct result of the economic conditions under which the workers lived. Great Southern Railways brought Daniel Flynn of Tipperary, a labourer at Ardnacrusha, before Justice Troy at Tipperary District Court for travelling from Oola to Limerick on 10 October 1927, without a ticket worth 2s. 8d. He explained to the court that while working on the Scheme he did not earn enough to buy a ticket. He made £1 from a three-day week in bad weather. He sent 5s. a week to his mother. After paying for his digs he had nothing left. He was given a 5s. fine with 20s. costs.[14]

The 'want of money' got the better of two entrepreneurial types in Clonlara in January 1929. A coiners' den was discovered in a hut there and two men were charged with producing and passing counterfeit coins. One has to admire not just their ingenuity but also their sense of timing, as the new Irish coinage had just been introduced after a protracted and passionate controversy. As yet, people would not be all that familiar with the new coins and mistakes could easily be made. The coiners were making Irish and English coins and the newspaper reported that the 2s. 6d. piece was very popular![15]

With the 800-odd Germans controlling the 5000-strong Irish workforce, it was inevitable that racial friction would surface, but very little of it appeared in the courts. There was the odd case of German foremen being assaulted by Irish navvies and of misunderstandings arising from relatively thin communication because of language difficulties, but a case in Killaloe District Court in April 1928 showed admirably how to play the dumb foreigner when it was to your own

■ Redundant trains: the giant effort takes its toll on equipment.

advantage. The case concerned three Germans, one charged with being drunk and the other two representing themselves as bona fide travellers. One admitted to having 'a leedle English', but he managed to translate for the other two. The justice asked if the summonses were translated into German before being served. Superintendent Mooney replied that all aliens were supposed to know either Irish or English once they were admitted to the Free State and they had to sign their names in Irish or English on admission. How the proud owner of a name like Adolf Burkhardt or, better still, Waldemar Czapiewski would attempt to do that is still a real test of anyone's imagination. The gentleman charged with being drunk conveyed to the court that he only had taken 'a leedle' and that he had less English. By all accounts he did seem to have tremendous difficulty in following proceedings and kept repeating 'trunkenheit, nein.' When the judge fined him 5s. he produced two half-crowns immediately without the slightest hint from his amateur interpreter![16]

Language was a problem not only for the Germans, but even for the Irish. The men from west Galway and from the Aran Islands stuck to themselves and conducted their business though the first language, much to the annoyance of some other workmen. The first Connemaraman to be brought to court was Edward O'Loughlin who was charged with stealing 30s. from a colleague and for assaulting him. The judge commented on the fact that O'Loughlin was the first Irish speaker to come before him and asked the garda to translate from the Irish. O'Loughlin was fined £4 for stealing and 5s. for assault.[17]

The fact that the Connemaramen did not mix much and were unable to speak English made them the butt of many jokes. Eventually the men from the west got

■ President Cosgrave with distinguished visitors and staff at Ardnacrusha camp. (Names on left.)

Pioneers At Ardnacrusha

Mr. E. Warbrook, System Operation Department in a letter says that while he cannot claim to be a pioneer he was at Ardnacrusha from 1938 to 1945 and recognised the photograph which we published in last months issue. He gives the names as front row from left — Professor F. Rishworth, P. McGilligan, T.D., Minister for Industry and Commerce, M. Deprevost, French Commercial Attache, Dr. T. A. McLoughlin, W. T. Cosgrave, T.D., President of the Executive Council, John F. Murphy, First Chairman, E.S.B., H. Eriksson, Swedish Consul, I. R. A. W. Weenink, Netherlands, Consul General Julian F. Harrington, American Vice-Consul. Back row (from left): Senor Ugo Aveline, Chancellor Italian Embassy, Frank W. Fraser, Canadian Government, Trade Commissioner, Col. J. O'Reilly, Aide de Camp to the President, M. Fernand Justice, Belgian Attache, Dr. Henry Kennedy, J. O'Farrell, P. Tierney, P. H. Egan, E.S.B. member, P. J. Dempsey, B. O'Sullivan, E. A. Lawlor, Herr Bach, Civil Engineer, Siemens Bauunion.

tired of being ridiculed, so, on Sunday night, 4 September 1927, they decided to put manners on their English-speaking colleagues by taking the law into their own hands. Being well fortified after some serious drinking in Clonlara and led by John MacDonagh of Lettermore, about forty of them went on the rampage. A bottle was sent flying through a window and that signalled the start of the proceedings. The MacDonagh brothers, the Flahertys and the Mannions, armed with sticks, stones and other missiles, attacked the occupants of the huts. They smashed everything before them and when the dust settled two people had to be hospitalized and a large number were treated for minor injuries. The gardaí from Clonlara were unable to restore order on their own, so their colleagues from O'Brien's Bridge were drafted in as reinforcements. Peace was eventually restored, but not before fourteen men had been arrested and transported to Limerick County jail in an open truck reminiscent of Black and Tan times. When the case subsequently came to court evidence was given by other workers against the attackers, to the effect that the Connemaramen did a lot of overtime and that this caused jealousy. Others stated that they were an uncivilized lot, not just because of their lack of English, but because they were a dirty crowd as they never washed! Both MacDonaghs were fined £2 and the others were fined £1 each for causing bodily injury and damage to property.[18]

Another Galwayman, with a German-sounding name, Josef Bulistron, was in the wars in January 1929 for slashing the faces of two gardaí with a razor outside of Stritch's pub in Clonlara. He worked on the Scheme and had been discharged from a mental institution in the previous November. He spent some years in the United States, but was deported from there when he was discovered to be insane. Strains of Ellis Island, but how times have changed.[19]

The various bus companies bringing men from and to the city were also frequent visitors to the courts. The most common charge was that of over-crowding. For example, a fine of 7s. 6d. was imposed on Denis Humphreys, a conductor with the Irish Omnibus Company for having 40 passengers on a 26-seater bus. Michael Leo, a conductor with Flannery buses, appeared in court on the same occasion in January 1929, for two similar offences.[20]

All news was pushed into second place by an horrific murder that took place in the days leading up to Christmas 1928. This became known as The Parteen Murder and it cast a dreadful pall over the whole area for the festive season and for the early months of the New Year. A German foreman, Jacob Kunz from Bavaria, was struck by an assailant with an iron bar at Parteen-a-lax while returning to Limerick. He died a short time afterwards. John Cox, an ex-soldier from Limerick, who formerly worked with Kunz, was later charged with the murder and robbery of £80 10s. from the Bavarian's pocket and of £409 10s. that he had sewn on his vest. The accused was charged in Limerick District Court on Saturday, 29 December and was remanded until 15 January 1929. In the meantime the money was found under a stone in Corbally. The case was heard in Dublin Central Criminal Court on 11 March. The trial lasted four days and the jury returned a guilty verdict. He was sentenced to hang on 11 April. The sentence was appealed and at the same time his family conducted a strenuous campaign to have him reprieved. But it was all to no avail; the sentence was made to stand. Cox was hanged on 25 April 1929, just two months before the official opening of the Scheme.[21]

■ The tailrace cut with Power Station nearing completion in background.

An interesting outcome of the case was that in the Court of Criminal appeal, when the hearing of the application for leave to appeal was made, the Chief Justice, referring to the matter of police interrogation and of taking statements from suspects, said that the fundamental principle is that no one is to be compelled to incriminate himself. With reference to a remark of counsel for the Attorney General that detection was the province of the police, Justice Hanna said that it was true that much crime might go unpunished if statements taken by the police were not made available, but that in the interest of the accused there should be a peace commissioner or some responsible person present during any interrogation or when he was required to make a statement. His remarks qualified for the next issue of *The Irish Law Times* and, in retrospect, not without reason.[22]

9 | Feeling Hard Done By

John Creagh was one of the busiest men on the Shannon Scheme. He was the manager of the Guide Bureau attached to the works. It was set up in June 1928 in response to popular demand. In its first two months of operation it conducted 10,000 people through the various building sites from Parteen-Villa (the weir) to Parteen-a-lax (the tailrace). As many as 2600 visitors had to be bussed and guided through on one Sunday in July. That was some achievement for Creagh and his team. People travelled from not only all over Ireland to see what was called the 'eight wonder of the world' but they also came from as far away as South Africa, India, New Zealand, the United States, Canada and the UK. Even John McCormack, who was forced to return suddenly from a singing tour of America because his wife had a horse-riding accident in Monasterevin, visited Ardnacrusha as he passed through Limerick. Before it opened officially the Shannon Scheme was big business. In September 1928, visitor numbers peaked with 32,776 travelling there.[1]

While visitors marvelled and guides expounded, while government ministers trumpeted and engineers quietly went about their business with increasing confidence, there were those who wished that Tommy McLaughlin and Paddy McGilligan had never come near their front door. These were the local people for whom the Shannon Scheme meant disaster – destruction of their homes and

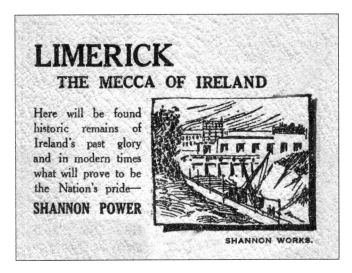

LIMERICK
THE MECCA OF IRELAND

Here will be found historic remains of Ireland's past glory and in modern times what will prove to be the Nation's pride—

SHANNON POWER

SHANNON WORKS.

■ Cover of tourist brochure for visitors to Ardnacrusha.

farms, ruination of their businesses and livelihoods, disruption of their social networks and relationships, loss of good health and, in some cases, loss of a parent, a son or daughter who met their death during the construction period. This was severe physical and emotional dislocation, and the huge personal investment exacted from all those people, young and old, living along the length of the project from Killaloe to Longpavement was seldom referred to.

Understandably, most of the talk and publicity was about progress and the future – what life would be like when the Power Station was up and going. There were few champions for the locals and their losses. Occasionally issues were raised in the Dáil or at County Council meetings but, by and large, it was up to them to make their own case. Some did and fortunately their correspondence still survives.

One of the earliest indicators for the general public that things were about to happen were the notices warning that some roads in the district would shortly be closed permanently. For example, two roads leading out of Ardnacrusha (now Parteen village), one going to Castlebank and the other going to Blackwater via Ballykeelaun, were to be cut forthwith. The famed O'Donoughue's Cross was to be no more. There were other roads similarly affected but it was the closure of the small road to Castlebank, Ballyglass and Roo East that sparked a row that was to last even for years after the opening of the Power Station.

The imbroglio concerned religion and education. This placed Fr John Moloney, the parish priest and manager of the local national schools, at the heart of it. For those who knew him, Fr Johnny was a character – formidably old-style – deter-minedly combative, fearlessly outspoken, highly protective of his people, and 'angular', even when trying to be pleasant. The immediate effect of the road closure was that his parishioners were cut off from their church in Ardnacrusha. Besides, excavations at the tailrace meant that the old Mass track through Boyle's land was also ruptured. People now had to travel up the canal bank to Blackwater Bridge and from there walk to the church, adding anything up to three or four miles to their journey. For those who lived below Davidson's pub a footbridge was provided.

School children were also badly affected, as their school was then situated at Garraun, just below where the tailrace bridge was to be built.

Initially, local politicians tried unsuccessfully to deal with the problem, but eventually, Fr John entered the fray. He had been preoccupied by another issue, viz. securing adequate compensation from the government for the school and its lands which were needed for the tailrace. He quickly secured a meeting with the Minister for Industry and Commerce and, among other issues, raised two points:

1. the question of the condition of Castlepark and Quinpool roads which necessitated his making a detour of eight miles to get to Castlepark church;

2. the need for building a footbridge near the Power House to enable his parishioners to attend church.[2]

He left the meeting under the impression that the Minister guaranteed to him that a bridge would be provided at the Power Station to shorten the route and he conveyed as much to his parishioners. If the guarantee was not carried out, he warned that it would have a bad political effect for this government, as his people were already greatly exasperated. Apparently some had attempted to claim compensation for the disruption to their lives, but the government was of the opinion that in law compensation was not payable for such inconvenience.[3] At any rate, nothing happened and, in classic style, it took three years before clarification emerged that there was a marked difference of opinion between the parish priest and the minister as to their respective interpretation of this meeting. The former claimed that he was promised a footbridge; the latter stated that at no time did he make such a promise. In a letter to John Nolan TD in 1931, he said that he had written to Fr Moloney in 1928: 'that there would be a service footbridge (i.e. a bridge for the use of the staff employed at the Power House) and that this bridge may later be used by Fr Moloney's parishioners.'[4] Curiously, he also mentioned that the matter was now out of his hands. Nolan passed a copy of the minister's letter to Fr John, who decided to write to President Cosgrave on the matter, enclosing Paddy McGilligan's letter to him after his visit of three years before. He repeated his case:

> These people are deprived of the facilities they had from time immemorial of going to Mass, sacraments and visiting their church. Now their children will have an extra burden of from three to four miles a day or seven to eight hundred miles a year put on them going to their schools – a great burden and expense. Now if this was a temporary thing it may be suffered but to be placed on them in perpetuity – why the Canonical finances of the parish of forty years and forty quarantines pale into insignificance before it![5]

The letter was passed on to Industry and Commerce who informed the President's Department that the Bishop of Limerick was now involved. This is what the minister meant when he said that the matter was now out of his hands. He advised: 'In reply to Fr Moloney who, by the way, is a most difficult man, it would be best that you should say nothing other than that the President will consult with the Minister for Industry and Commerce on the matter at an early opportunity.'[6]

Apparently, there were two proposals on the table. One was to allow the parishioners to use the service footbridge at the Power House, but this was considered unsafe. The second was to build a brand new footbridge at considerable cost. Seemingly, the minister was not averse to building a dedicated footbridge, had it been essential and justified. But as he pointed out to Fr Michael Maloney, Limerick's Diocesan Secretary, 'I don't want to incur any expenditure in connection with the Shannon Scheme other than is absolutely necessary….' Besides, he was worried that if he had to build this bridge it could very well prove a precedent for other claims for bridges along the canal on grounds such as access for cattle to watering places, facilities as between suppliers of creameries, etc.[7]

As it happened, the bishop consulted Fr John as to how many of his parishioners were disconnected by the canal and concluded that the expense of a footbridge would not be justified, as there were so few people involved. His opinion was communicated to the minister in June 1931. There the matter rested.[8]

Fr John's reputation as a difficult man developed really during his negotiations

■ Waiting for the waters to arrive in the headrace.

with the government on the question of compensation for the school which was eventually knocked down. Fr Hogan, his predecessor, was much more amenable and presented no great challenge to government valuation experts or to ministers. Paddy McGilligan did admit that in his meeting with Fr John he found the latter to be totally intractable and could not find the slightest chink in his armour.[9] In fact, while he was the person conducting the meeting, he found himself suffering a severe grilling from the parish priest!

The school was closed earlier than other schools, on 4 July 1927.[10] This was as a result of constant pressure from Siemens Bauunion, who were afraid of falling behind in the contract. Already they had undertaken blasting operations in the area when children were not present, but there was a danger to the school equipment and furniture. The manager, Fr John, was under pressure to provide alternative accommodation for the children, but he could not build until he got compensation funds for the old school. Besides, even he was unsure as to what was happening and relayed to Professor Rishworth the rumour in the parish that the school would be closed down permanently.[11] As a result children were also leaving the school and enrolling in town schools.

Eventually Industry and Commerce decided that something had to be done quickly. It contacted the Department of Defence to see if an army unit was available which could be used as temporary accommodation for the children.

The army did have such a unit, but it was based in the Curragh. However, it was quickly transported to Ardnacrusha village and erected in a field owned by the parish priest. Made of corrugated iron it was lined with matchboard sheeting.[12] There was an irony here in that this hut came from a place called Tintown Camp in the Curragh and was being erected in a spot not far from where hundreds of Shannon Scheme workers lived in makeshift huts known locally as Tarbarrel Lane!

The rapid evacuation of the old school and its adjoining buildings in the summer of '27 meant that the two teachers, Kieran Kelly and Johanna Ross, had to find alternative accommodation. Fr John negotiated with Arthur Taylor, the government valuer, and secured a sum of £100 cash for both teachers for eighteen months, in lieu of schoolhouses. Both teachers accepted and in his letter of acceptance to Taylor the manager warned that more money would be expected if the new residences were not ready within the eighteen months.[13] Taylor advised both teachers that the money would be through in two weeks. Three months later Kelly was writing to the Department of Industry and Commerce reminding them of Taylor's grant which had not arrived and that now he had to pay rent in Limerick far in excess of the grant promised. He was angry: 'I now make a formal application for same and I hope it will be paid immediately. Otherwise I shall be compelled to adopt other means for its recovery.'[14]

Ominously for Kelly, Mrs Ross' cheque arrived in November with a receipt that she was to sign accepting £100 'in full as compensation for loss of residence'. As this was not what was agreed between the assessor and her manager, she sent the receipt back for amendment, along with the cheque. Eight months later she had heard nothing.[15]

If negotiations on the school were anything to go by, it was going to be some time before she and Kelly were to be paid, if ever. The Honan Trust, of which the bishop and the parish priest were the trustees, legally owned the old school complex, even though Kelly was claiming that he owned some of the land, a matter which put him under suspicion with officials in the Department of Industry and Commerce. Working through two sets of solicitors and valuers, the church and the government eventually arrived at an agreed figure of £929 for the school. Arthur Taylor then suggested that the cost of the temporary schoolhouse should be deducted but Fr John went apoplectic and the assessor quickly withdrew his proposal. Even a year later Taylor had not forgotten Fr John's terrifying performance and wrote to the Department of Industry and Commerce about it:

Although the matter was approached with all possible tact the Reverend Gentleman launched out into an attack on the Minister, and said the government did not know who were their friends. I pointed out that I was not in any way concerned with politics and that the suggestion made by the Department of Finance appeared to me to be perfectly proper and reasonable and there was nothing in the matter except business pure and simple. The present structure was worth something, the department was willing to let Fr Moloney have the structure at its present value, and that it might be useful as a parochial hall and so on. Fr Moloney stated that the structure was useless, simply an old army hut that could not be sold, and in order to protect the government from their enemies who sat on the Clare Board of Health he had to have a man employed regularly on the closets, which were in disgraceful condition, and that it was only a dispensation of Providence that there hadn't been an outbreak of fever … In the end he calmed down but was full of grievances at the way he and his people had been treated … He told me plainly that he would not give three halfpence for the structure and to take it away and do what I like with it. He is an extraordinary man … I do care not to criticise any clergyman, but I should prefer not to be instructed to interview Fr Moloney on this matter again.

It is obvious from this letter that the poor man, like the minister before him, had been badly bruised by the encounter and he advised the Department to pay the £929 compensation immediately. He added that the minister should consider himself fortunate to get away so lightly as the new school buildings were to cost £2,200![16]

Protracted negotiations and inordinate delays seemed to be a feature of compensation procedures. This may have been due to the fact that a number of government departments were involved but it certainly made life almost impossible for some people. Ms Helen Ryan of Ballinagarde in Co. Limerick was a case in point.

Ms Ryan's mother, Elisabeth Ryan of the same address, owned a stone two-storey building and land just below the village of Ardnacrusha. Helen managed the property for her. The Board of Works took a lease on them early on in the Shannon Scheme and housed a number of guards there. The house is now gone, but it was known afterwards as 'the old barracks'. It was situated 150 metres from the centre of the tailrace, which meant that it was sorely affected by the blasting operations. Flying rocks had damaged the roof and earth tremors had caused some of the walls to bulge. So the guards for their own safety had to move out of the house and occupy nearby huts owned by Ms Ryan, for which she charged an annual rent of £10.[17] As a result, the Board of Works decided in mid-1928 to negotiate with Ms. Ryan for a release from their tenancy on the best possible

■ Giant flash: first test on the earthing system at Ardnacrusha.

terms. The house was held under a yearly agreement at a rent of £52 per annum, inclusive of rates but exclusive of maintenance.[18]

In October 1928 Ms Ryan wrote to Paddy McGilligan listing the damage to the house and lands that had been drained of well water because of the deep excavation of the tailrace. She claimed that both of them were now derelict and useless. 'I wish to draw your attention to the fact that my way of living is completely gone from myself and my mother and I shall be grateful if the matter is looked into as soon as possible, as we could not live without rent of house and lands.'[19]

Unfortunately for Ms Ryan and her mother, she had to make do without the rent for another year; in August 1929 she was writing again to the minister pleading once more for payment. Yet another letter followed from her in October and a third in November. But no money came from the Department.

Apparently what was happening in the background was that the Office of Public Works all the while was trying unsuccessfully to get compensation from Siemens Bauunion for damage to the house and the cost of rent for the plot.

In January 1930 Public Works gave up on the German company, claiming that they never replied to correspondence.[20] Arthur Cox, legal advisor to Siemens, then became involved, listing Siemens Bauunion's letters to Public Works and others on the matter and stating that his company always repaired damages due to flying stones such as broken window panes. 'My clients refuse however any responsibility for damages due to vibrations … which damages could not have been avoided by any protection whatsoever.'[21] This firmly put the ball back into the Board of Works court which was now trying to extract the money from Industry and Commerce.

Professor Rishworth was then called on for an opinion and he advised Industry and Commerce that Ms Ryan's claim seemed to depend on the elucidation of a legal point relating to the damage caused by atmospheric vibration due to the blasting.[22] There was no way she could fund an expensive legal challenge particularly if she was as short of money as she claimed.

The last entry in the relevant file is another written appeal from her for payment on 30 January 1932. She said that her brother would not accommodate herself and her mother after 1 April and would be grateful to the Minister if he could find a house for her in a select locality. Apparently Clare County Council was demanding rates from her for the house and lands and she claimed that she was not in a position to pay. She also claimed that she could not even settle her other personal accounts due to shortage of cash. Unfortunately, there is no evidence of her ever being paid what she claimed but she certainly deserved it, and much more, for the emotional wear and tear she had to endure for five years.

The Browne family from the village subsequently bought the old barracks and land, but they too had their own problems with claims for damages.

In May 1930 Mrs D. Browne wrote to the government:

Sir,
Having made many applications to have my claims settled (getting no reply) I wish to bring under your notice the six houses which were built in Ardnacrusha-Parteen that have been and are idle for the past six months owing to the blasting operations, which for three and a half years were in such close proximity to us, also our own residence and all the little village which is shook asunder. As well our land was not settled for as we naturally wanted water which was taken away from us. Should the Government kindly settle our claims and presently give us a share to carry on with for the repairing of the houses as our own little spare capital we invested in Great Southern and Western Railways shares which are paying nothing. I remain Sir,
    Yours obediently
    Mrs D. Browne.[23]

■ John Browne of Parteen, whose houses were damaged by the blasting in the tailrace, outside Cruise's Royal Hotel in Limerick.

The family ran (and still run) the public house in the village and had built the six houses on the site of the old RIC Barracks that was burnt down during the War of Independence. They had German tenants in the houses during the early days of the Scheme, but these had to be vacated because of the blasting nearby.

The earlier claims referred to in Mrs Browne's letter relate to applications for compensation of £1000 for their houses and lands. Nearly four acres of meadow were covered with stones so that no hay could be saved from 1926 to 1929, losing them an estimated £20 per annum. Her solicitor, T.O'Brien. Kelly, put in his first claim for compensation in December 1929, again in 1930, followed by Mrs Browne's letter in the same year. It was over a year later that John Browne managed to get a meeting with the secretary to Shannon Power Development, V.M. MacMahon, on 23 October 1931.

The minute sheets of the meeting, which lasted over two hours, are most revealing. Browne began by saying that he had been in Limerick Court when judgment was given in the cases of *Siemens Bauunion v. Bourke and Campbell*. These were interesting cases taken by Synan Campbell and John Bourke, both from Clonlara, who sought compensation for damages from the contractors as a result of the blasting. The latter was claiming £169 17s. 6d., and the former £178 13s. Both cases were running in tandem with that of Patrick Boyle who also claimed damages for injury to property because of the blasting. All three were listed for hearing on 17 July 1930. Boyle settled out of court but the cases of the other two were adjourned until November, when both were awarded £30 and the costs were set at High Court scales. However, a stay was made until 2 March 1931, when the judge announced that in view of a House of Lords decision in a similar case he had no option but to overturn the earlier decision and decide in favour of the contractors. The Lords, by a majority of one, had held that liability for compensation for air-vibration set up by blasting could not be proved. As a result, until this legal decision was altered, it governed the courts and dictated the attitude of the Department of Industry and Commerce's attitude. This suited the government just fine, as it did not have to pay out compensation to claimants. However, the judgment must have been very difficult for plaintiffs to comprehend when they surveyed their properties after blasting operations when walls which were fine before suddenly began to bulge or crack noticeably. At any rate, the official informed Browne that, because of the House of Lords decision, there was nothing that the minister could do for his houses in Ardnacrusha. He advised that if Browne submitted a detailed claim in writing, he would give a written reply. The same advice about a written submission applied to the land acquired from him permanently and, if followed, the matter could be put in train for arbitration. There was no way, however, that arbitration could apply to the houses as this question lay outside the scope of the Act thereby excluding the arbitrator from hearing the case, the merits of which had already been decided by the courts.

In some private notes relating to the meeting the official had registered that Browne had already received compensation for the land acquired temporarily and for subsequent damage to it. It was noted also that he had refused a generous offer of compensation for lands acquired from him permanently and that the case would be referred for arbitration. These negotiations had earlier broken down on the question of compensation for loss of work. The arbitrator apparently had stood firm and was relieved when Browne refused the offer made. The former was of the opinion that the latter had a claim but that it was grossly exaggerated. The notes went on about the above-mentioned Clonlara cases when the presiding Judge McElligott invited the contractors and the plaintiffs to appeal against his decision. The contractors would not, because it was in their favour; the plaintiffs could not, because of the risk of being swamped in costs if they lost. The fact that costs had been set at High Court scales was a deterrent in itself for most people. The notes registered that Browne's claim was so large that it would have to be heard in the High Court, but there was nothing to prevent him from taking it there.[24] Nothing, except lots of money. In the event, the case eventually went for arbitration and in the absence of any further relevant documentation one can only assume that the settlement made favoured the government rather than the claimant.

The conclusion of the *aide-memoire* was interesting. 'We cannot admit liabil-

■ The intake area where the waters of the headrace enter the station.

ity. If we did the Minister would have to pay compensation for every crack in every house for miles around. There is no doubt, however, that some people have suffered from the blasting… Until the legal decision is upset it holds the boards.'[25]

It has to be noted, that this meeting happened at the end of 1931, nearly seven years after the commencement of the Scheme, when Browne's lands, along with those of many other farmers, were acquired either temporarily or permanently, by the government. It was only now that he was getting advice on how to proceed with his application. As in the Ryan case, state bureaucracy was in no rush to pay out money and, while the above quote would suggest that officials recognized the personal suffering of people caused by the explosions, personal hardship was not an element readily factored into the arbitration process, if at all. For a relatively young administration, the system seemed to be particularly muscle-bound or suffering from counter-checks within itself, which may just underline the conservative nature of those operating it. This is not to say that all those involved in state machinery were lacking in humanity. Arthur Taylor, the state assessor, for example, comes through his various and extensive reports as a most delightful man, scrupulously thorough and direct in his dealings with individuals, claimants and civil servants alike, sympathetic too towards genuinely aggrieved parties, and all the time, a most shrewd reader of human nature, enabling him to detect (and deliberately note) the smallest scintilla of venality generally endemic in any such process and procedure involving compensation. The fact that the contractors lost money on the Shannon Scheme and that the government was strapped for cash ensured that both sides developed systems that got maximum value for the very minimum of expenditure. The public face of such systems frequently appeared harsh and exacting.

Perhaps one of the more celebrated compensation cases was that of Mrs Catherine Davidson who lived just below the tailrace bridge at Parteen. Her home and licensed premises were within the danger zone affected by the extensive blasting operations of the canal. In fact, the house and pub were very close to the centre line of the rock cut making it extremely vulnerable to vibrations and flying rock.

On 8 June 1927 Mrs Davidson wrote to the Resident Engineer at the Strand Barracks asking about protective measures for her property against the blasting operations.

> I trust that some protection will be afforded me without delay as already the premises have been subject to some damage as a result of the explosions. I am also informed that the road passing by my premises where I carry on a licensed trade will be closed during the period of the construction of the tailrace. This will, as you can understand, materially damage my trade owing to the fact that the public will then no longer use the said road.[26]

As a result of her request, the contractors shored up the premises with props, put up window guards and covered the slate roof with a wooden roof. Apparently these measures proved ineffective and, in spite of the timber covering, flying rocks damaged the roof and large cracks appeared in the walls.

A tragic occurrence took place on 17 July, when eighty-six-year-old Mrs Flynn, Mrs Davidson's mother, was opening one of the timber window shutters. It fell from its place crushing her against the parapet wall at the front of the building. She died shortly afterwards. The inquest, at which the minister was represented, expressed the opinion that there was gross negligence in the putting up of the window shutter. Her accident caused all blasting in the area to cease temporarily. Once her body was removed for inquest, however, blasting operations were resumed again.

The death of her mother, the continuing damage to her house and the effect of the excessive explosions, all combined to affect Mrs Davidson's nerves. It became imperative that Mrs Davidson and her son and daughter should vacate the house immediately.

Even though a temporary footbridge to facilitate those who wished to travel to the village was erected, the closure of the road had badly affected her trade in the pub thereby eliminating any residual reason for her to stay.

Mrs Davidson's solicitors, Connolly and Co. of Limerick, then wrote to the Minister of Industry and Commerce, shortly after the accident to her mother, in order to expedite matters. Interestingly enough, they mentioned that their client kept occasional lodgers and supplied lunches to the Irish engineering staff who were engaged in supervising the work of the contractors. All that was now finished, the letter went on, and of late people are prevented from passing along owing to the danger of being struck by flying rocks. Apparently, Mrs Davidson had to leave the house on two occasions recently because of the proximity of the blasting.[27]

Arthur Taylor, the assessor, was then called in and he confirmed that

> since the death of her mother… Mrs Davidson is in an extremely nervous condition. The blasting is on a colossal scale and every now and then Mrs Davidson has to leave her home and go down to a place of safety together with any customers that may be in the bar at the time… I suggest that it would be advisable to close the premises and to offer her compensation for finding accommodation elsewhere until such time as no further damage is likely to arise from the blasting operations.[28]

Taylor's corroboratory report was enough to move matters forward and he was instructed to enter into negotiations with Mrs Davidson with a view towards a settlement. By early October he reported that he had reached a most satisfactory outcome from the point of view of the Department of Industry and Commerce. He had examined Mrs Davidson's trading books and concluded that her net profit for a week was £6. Considering that she would have to secure accommodation for herself and her two children in Limerick and to travel by car to inspect the property regularly, he would recommend that she be paid compensation of £6 per week.[29]

The Department of Finance sanctioned this proposal, but not before checking with the assessor as to whether Mrs Davidson's tax returns were in order! J.J. McElligott suggested that the local inspector of taxes would be willing to furnish this information if Mrs Davidson gave her written permission.[30] This exercise took four weeks, which is a long time for someone in a highly nervous condition, whose house was being literally bombarded on a daily basis and who had a business to run and two young children to care for. At any rate Mrs Davidson's tax affairs were in order, so the payments were made from January 1928, even though she was forced to vacate the house in early December because of its dangerous condition. These payments would last for a period of twelve to eighteen months.[31]

■ The enormous steel grill gates preventing debris from entering the turbines.

Work continued on the tailrace and by early summer 1928 J.K Prendergast, the Resident Engineer, was of the opinion that unless Mrs Davidson's house was properly shored up that it was liable to collapse at any moment. The fact that all of her furniture and effects were still in the house, some of which had been damaged by rain pouring through extensive holes in the roof, could lead to a very expensive claim. Prendergast also pointed out that the contractors were not taking the matter very seriously.[32] Further devaluation of Mrs Davidson's property took place when it was decided to build the tailrace bridge at its present location, and not farther down the road towards Limerick. This meant that part of her car-parking area in the front of the house was truncated, bringing her considerably closer to the road. At the back of the house, the railway line to Ardnacrusha cut through an accommodation paddock that she and her predecessors had rented from the Glosters for over eighty years. All in all, with the top floor of the house having to be rebuilt completely and with the general despoliation of the property outside, Mrs Davidson was entitled to some compensation.

In mid-1929 she and her engineer W.J. Holmes submitted plans to Siemens Bauunion for the rebuilding and general refurbishment of the house with an estimate of £750.[33] The contractors stalled immediately and said that refurbishment was not their responsibility as the damage was caused by vibrations. They were now using the House of Lords agreement that the government had used against John Browne. 'We could only acknowledge liability if the decision were made in court', they emphasized.[34] However, they did concede that while they did not have a legal liability, they would be prepared to pay, without prejudice, half the sum in question, i.e. £375, in settlement of their liabilities.

It would have been interesting to see what the government would have done had the Germans played hardball and not made the part-payment concession. There would have been a protracted stand-off involving lengthy legal proceedings. The contractors' contention was that they were responsible for such damage done by blasting as could be proved to be due to negligence. The government, on the other hand, relied on a clause in the Conditions of Contract that placed liability on the contractor for all damage to property arising out of the blasting operations. The House of Lords judgment though effectively modified this clause somewhat.

A summary of Mrs Davidson's case for the Chief State Solicitor contained one very interesting and little-known fact:

> It is understood that the contractor is insured against all compensation claims arising out of the blasting operations and that in repudiating responsibility he is acting on the advice of the insurance company concerned which is located in Germany. There is, however, no official knowledge of this.[35]

In the meantime Mrs Davidson's claim had risen to £1000 because of interest costs, architect's fees, etc. Arthur Taylor reported that the claim was reasonable in the circumstances: 'However, I queried the claim and fought it on at least six occasions with Mrs Davidson and her Valuer and Engineer. Eventually I am happy to report I succeeded in reaching a most satisfactory settlement, namely £800 in full satisfaction of the claim...'[36]

On 17 July 1930 the Department of Finance finally signed off on Mrs. Davidson's claim, allowing her £800 in respect of compensation, 25 guineas Valuer's fee, £6 per week for accommodation until the end of June, and a final lump sum of £54 for alternative accommodation, as her house would not be completed until 30 August.[37] This brought to a close what must have been a nightmare for Mrs Davidson and her young family. The engagement with the various government departments was a veritable war of attrition, on top of the trauma of being blasted out of one's home and losing a loved one in the process. Happily, Mrs Davidson and her family lived in their house for many years after that and provided a welcome service for all bona fides and even for those who may not have been as eminently qualified!

While the issues involved in compensation cases were complex thereby delaying an outcome, there were some instances when matters were decided relatively quickly. William O'Grady's case was one of those. His farm was in Blackwater, where nearly thirty acres of land were flooded by the contractors in 1925 as a result of the canal cutting. For four years the land was covered in water preventing him from sowing crops or making hay on it. In February 1929 a claim was made in the courts for £200 19s. 3d. and costs against Siemens Bauunion. It quickly transpired that the contractors were not responsible for the damage so that it was up to the government to settle the matter. Arthur Taylor then got involved and a conference was arranged between the various parties.[38] After intensive arguments he got the claimants to accept a sum of £130 in full satisfaction of their claim including costs. Taylor recommended to the Department of Industry and Commerce that it should accept this settlement and pay up immediately.[39] By early August Finance had approved the payment.[40] The whole process had taken only six months, which was comparatively speedy, but there were no great complications in the case.

There were many people adversely affected by the construction works who could not afford to travel the legal route to progress their grievances. There were many others who, as we know from the numerous inquests, could not write or even sign their name so that they could not correspond with the government or state their case. Their stories will never be known, which, unfortunately, is our loss. For those people, unrecorded and now forgotten, as well as those of whom there is some evidence of what they went through, the cost of progress was high. The minimalist awards were of little comfort for what they and their families suffered and lost. For them it was a huge investment in the future and in the next generation.

10 | The Battle of the Tailrace

The year 1932 was a memorable one. February signalled the end of ten-year-old Cosgrave administration. March saw the formation of a new government under de Valera. April brought the worst employment figures since the depression began, with over 31,000 out of work. June was marked by the celebration of the fifteen-hundredth anniversary of St Patrick's coming to Ireland and the Eucharistic Congress. But it was July, perhaps, that was best remembered for years after by the majority of Limerick people, and by the Abbey fishermen in particular. During the second week of July the Abbey fishermen took on the army, police and water bailiffs in a series of engagements that have since been known as the Battle of the Tailrace.

For centuries the Abbey fishermen and the river Shannon were practically synonymous. For hundreds of years the fishermen plied their narrow fishing cots between the city and the mouth of the Blackwater river on one side, and up to Illaunareum Trench on the other, where they exercised their ancient skill of the snap net. During the Williamite siege of Limerick the fishermen were, reportedly, given special privileges. By the 1700s they were recognized as an incorporated body. And by 1905 the forty-odd fishermen, drawn from eight to ten families, emulated the Limerick tradesmen of previous centuries and took to themselves the rather grandiose title, 'The Abbey Guild of Fishermen'. But in the 1920s the

■ The Abbey Fishermen poling their cots up a difficult stretch of river.

death-knell was sounded for their salmon industry, worth an estimated £20,000 annually.[1] The building of the Shannon Scheme was to destroy their main source of livelihood and to terminate their long tradition of association with the river.

Fears for salmon fishing on the river were generated among the Abbey fishermen with the publication of the plans in the mid-1920s for the electrification of the Shannon. The river was to be dammed at O'Brien's Bridge and diverted into a headrace with a hundred foot fall into the tail-race before returning to the main river below the Lax Weir at Parteen. This plan meant that the main volume of water from the Shannon would be siphoned-off, creating another river with a much stronger current that, in effect, would destroy the Lax Weir and other fisheries above it as far as Killaloe. Up-running fish would not be able to pass from the tailrace to their spawning grounds by the power station, and those returning to the sea would be trapped in the headrace at the intake gates.

Two other factors increased apprehension among fishermen. Firstly, the Minister for Industry and Commerce, Patrick McGilligan, bluntly stated in 1925 that the White Paper on the Shannon Scheme 'definitely foreshadows that it might be necessary to neglect the fishing interests in the cause of power production. It foreshadows that fishing interests will not be allowed to predominate against the greater interests of power production.'[2] Secondly, a number of the best salmon waters in Europe and North America were being destroyed as a result of indus-trialization. Rivers were being harnessed for industry thereby preventing the salmon from reaching their normal spawning ground; also pollution resulting from industrialization was simply decimating stocks. Ireland's rivers had suffered relatively little in this regard, but the next few years were to see that change.

■ Abbey fishermen returning with catch, seated and paddling.

The 1925 Shannon Electricity Act made the Minister for Industry and Commerce liable for any damages to Shannon fisheries due to the construction of the Shannon Scheme. This, however, was little consolation for the Abbey fishermen, who would be faced with the prospect of leaving their industry and being retrained for some other employment. Even at this stage the fishermen argued that money could not compensate them for what they would lose if measures were not taken to protect their industry.

The debate continued throughout the construction period of the power station. The owners of the fifty-odd stake weirs and approximately one hundred nets on the river continually lobbied TDs and assiduously made representations to government departments on the issue. Curiously enough, the year 1927, when building operations at Ardnacrusha were in full progress, was a record year for some of the fisheries. It was also the year when the Abbey fishermen consolidated their business by buying Eyre Powell's fishery, which they had worked for years, stretching for five miles from Corbally to Doonass on the south bank of the river.[3]

On Monday, 22 July 1929, President Cosgrave performed the official opening of the Shannon Scheme. For the following weeks water was allowed to trickle into the canal, slowly filling the headrace so that the banks would not give under the weight and volume of the water. By October electricity was looked upon as tangible proof of Ireland's independence and separate statehood. Less than two miles downstream the worst fears of the Abbey fishermen had been realized; their fishing grounds had been sacrificed for progress and they themselves had become the most recent victims of industrialization. From the very outset fishing was prohibited in the tailrace where hundreds of salmon were dying daily. It was

■ The Abbey fishermen were not allowed to fish in the tailrace.

alleged by *The Limerick Leader,* 'the high mortality amongst fish is due to pollution of water caused by the dumping of "spent" salmon or slats which are held in the headrace'. Spawning salmon were also entering the tailrace and were unable to travel farther upstream. Besides, the turbines mangled millions of salmon fry.[4] Various attempts were made to overcome these problems.[5] The idea of a ship-lift to accommodate the salmon at the intake dams proved to be impractical primarily because the fish would not face into still water. The Limerick Board of Conservators placed a barrier of electrified chains at the mouth of the tailrace so as to prevent the fish going upstream, but this too was unsuccessful.[6]

Cumann na nGaedheal left office in February 1932. By this stage the Abbey fishermen and their families were in a bad way, having been deprived of their livelihood for the past three years. Ironically, they still had their fishery rights. The arrival of the Fianna Fáil government brought fresh hope to the fishermen, particularly as the Shannon Scheme affected de Valera's constituency in Clare, and some of his constituents had made representations to him on the fisheries question. One newspaper reported:

■ The Lax Weir with motorized launch in foreground.

> The fishermen have repeatedly called on the Cosgrave Government to stop this
> state of affairs, and now expect the Government that has the lot of the plain peo-
> ple at heart to act immediately and to do something to save those fishermen and
> their wives and children from starvation.[7]

On Tuesday, 5 July 1932, in the Dáil, Limerick Deputy James Reidy asked the
Minister for Lands and Fisheries, Patrick Ruttledge, 'if he will state what steps are
being taken to prevent the destruction of salmon in the tailrace at Ardnacrusha,
and if the people engaged in the salmon-fishing industry whose means of livelihood
have been interfered with by the Shannon electrical development will be permitted to fish the tailrace and thus enable much valuable food to be saved from
destruction.' The minister replied:

> The staff of my department have been inquiring into a recent finding of dead
> salmon in the tailrace at Ardnacrusha and if the cause of mortality can be discov-
> ered I shall consult with the Minister for Industry and Commerce as to the steps
> to be taken to prevent its recurrence. I am not prepared to revoke the By-law pro-
> hibiting the use of nets in the tailrace.[8]

This reply seemed to have been the last straw for the fishermen. By the weekend
they had resolved to defy the ban on netting in the tailrace unless stopped by superior force. Their decision was conveyed to the Limerick Fishery Board of Conservators
on Monday, 11 July. The Board requested the fishermen to reconsider the decision but this was refused.[9] Positions had now been taken for what was to follow.

That night at 11.30 the full fleet of the Abbey fishermen numbering twenty-four boats, each containing two men, gathered at St Thomas' Island. On the stroke of midnight the fleet approached the tailrace. An official of the Fishery Board shouted to the fishermen from the bank that fishing inside a particular mark was prohibited. The warning went unheeded and the fishermen continued on their course. Bailiffs in three motorboats patrolled the mouth of the tailrace. Gardaí were also on duty in launches. Hundreds of onlookers lined the banks. Slowly the angling cots arrived and some of them nosed over towards the bailiffs hut to extend their line in order to make it more difficult for the bailiffs and police to prevent their entry to the canal. The turbines at Ardnacrusha had been turned on at full strength creating a tremendous current that made the handling of the light fishing boats more difficult.[10] William Lysaght describes the next few moments in his book, *The Abbey Fishermen*:

> …there was a moment of hesitation, but with a cry of 'Up Garryowen' and a few deft strokes of the paddles, Randy and Lully Hayes sent their boat surging in between the chains behind one of the launches. This was the signal for concerted action. In a matter of minutes all the boats were inside – the battle was on. Some boats made to go upstream towards the power station. The bailiffs followed. The crowed cheered the fishermen on. From Parteen Bridge stones were thrown at the boat injuring one bailiff, Thomas O'Connor, though not seriously. Nine nets were seized. Gardaí took the names of 42 men. Three boats, reportedly heavily laden with fish, escaped. Four or five shots were fired in the air to disperse the crowd. The last net was seized close to 3 a.m. and then the boats returned to their base at the Sandmall.[11]

On the following day, Tuesday, the fishermen planned their strategy for that night, and at 8 pm the full group assembled at the Sandmall. Almost immediately they dipped paddles and set out for the tailrace forty-five minutes away. Peadar O'Donnell, the former Sinn Fein TD for Donegal, was present that night as a reporter for *An Phoblacht*. He gives an eye-witness account of the events:

> I arrived at the tailrace one of the war evenings; a wet evening but crowds of Limerick folk were there waiting for the fishermen to come. Police in great numbers; a dark bundle of a dozen men over near a shed pointed out as the bailiffs. And a-lack-a-day, the man in charge of the bailiffs was one who had a great reputation among the Volunteers. I felt terrible ashamed for him as I drew near to the crowd of the bailiffs. I talked to him: what I said doesn't matter, I suppose. Then I did a meeting of the bailiffs. I asked them to go on strike for the night and I put my heart

■ The nerve-centre at Ardnacrusha. Control-room operators monitoring output of power throughout the country.

into the talk, while bewildered policemen with horned, stumpy necks pushed into the crowd. And I would recognise again one face full of enthusiasm when I seemed to be winning the bailiffs. But I lost.

At this point the fishermen appeared in view and approached the tailrace. The crowd cheered. Peadar O'Donnell continues his account:

The bailiffs went to their motor boats: the war was on. Just picture it: two powerful motorboats full of well-coated bailiffs wait while frail two-men shells of boats go quickly into the tailrace. Out go the nets. And the crowd cheers. When a salmon strikes the cheers become a roar. Suddenly the motors are set going and in come the bailiffs' boats. The scene suddenly becomes sickening; the motorboats crash in among the coracles and grappling hooks reach out. There is a crash and a man of the McNamara's is overboard; clothes, boots and all, he is down the mill-race of the flood. But the bailiffs hang onto their prey, the boat and the nets. McNamara swims ashore. The crowd rush to the water's edge and there is excitement.

Policemen with drawn batons move about, some imploring, some threatening. One policeman reminded me of Mickey James in Rat-Pit: I expected him to invite us to sniff the scent of dead men on his baton. Suddenly there is a stampede: military with fixed bayonets are clearing the embankment: police are using batons.

I picked up a docker out of the dirt of the road into which he had been hammered by police. I witnessed this incident from the start: the docker, another of the McNamara's, just would not run: he was sulky, grudging, but that was all. He was struck with a baton: I saw a policeman box him, another kicked him, kicked him heavily. A man in a raincoat – I was told he was a police officer – was in a group that hammered the man. I got the man from the police and I drove him into Limerick: his face was bleeding, his clothes were a mess. But the man was

■ President and Mrs Cosgrave and family at opening ceremony, July 1929.

quite gentle: he wouldn't run from the police. I saw a few other incidents that were shameful to the police, and in any case the man in the raincoat was there: policemen seem to think that the presence of an officer demands roughness from them: they reminded me of a buck-navvy gaffer when the travelling ganger appears. I saw incidents where guards were considerate.[12]

The curtain of darkness brought an end to the baton-charging and stone-throwing for one night. In all, seventeen boats and ten nets were seized; ten boats escaped; forty-eight names were taken. The confiscated nets and boats were put in a military lorry and brought to Sarsfield Barracks. During the night a party of military patrolled the banks of the tailrace.

Wednesday brought a flurry of activity at national level. The Minister for Lands and Fisheries, representatives of the Department of Industry and Commerce and local politicians met in Dublin to discuss the situation in an attempt to take some heat out of it. In the meantime the fishermen regrouped their depleted forces and prepared for that night's foray.

At 10.00 pm ten boats entered the tailrace. A huge crowd turned up to watch, even though the military were on duty. The boats were allowed to enter the tail-race but once there the bailiffs closed in. Shots rang out as the fishermen resisted. Pandemonium broke out on the banks once more as people stampeded. The military on duty with fixed bayonets ordered people back to a certain distance. By 3 am all boats were in the possession of the authorities.[13]

On Thursday it was rumoured that the fishermen were building rafts. Denis Hayes, a fishmonger from Roche's Street, Limerick, acting as intermediary for the fishermen, confirmed that the rumour was true. But he said that he would not be supplying them with nets because of the danger of a drowning tragedy. He also announced that the men would not be marching to the Employment Exchange that day to demand unemployment benefit, on the grounds that they could not now earn their livelihood; it was decided instead to wait the arrival of the local TDs from Dublin, where they had been negotiating with the Minister for Lands and Fisheries.[14]

Friday night saw the fishermen's final protest at the tailrace. Four men arrived with one-and-a-half-boats! The two men in the stern of the derelict craft were soon swamped; the other boat and net were confiscated. Again a huge crowd attended.

While these events were taking place on the river, a meeting of Limerick Corporation was unanimously adopting a resolution protesting against the naked militarism, reminiscent of the Black and Tans, carried out on the tailrace during the week. The excesses of the military and garda force also came in for condemnation at a Labour Party meeting.[15]

At any rate, as far as the fishermen were concerned the Battle of the Tailrace was over, but their campaign continued.

On Saturday the full group of fishermen attended a meeting in St Mary's Hall. Limerick Trades Council and the Labour Party were also represented. Deputies D. Bourke and J. Reidy reported on their negotiations with the Minister for Lands and Fisheries and said that the Minister proposed to close down the turbines during weekends with a view to releasing fish from the tailrace, and that he would be introducing a bill in the Dáil dealing with Shannon Fisheries during the autumn session. The fishermen met in closed session later in the evening and rejected the Minister's proposal. They decided that since they were not allowed to fish at weekends anyway, closing down the turbines then would be of little use to them.[16]

On the following Monday, 18 July, the fishermen had their boats and nets

■ A tour of the generator hall on opening day.

returned to them. Months later charges against the fishermen of illegal fishing were dropped.[17]

In spite of pressure from the fishermen, from Limerick and Clare TDs, from various bodies in the United States, and in spite of the government's awareness of the need for legislation, it was two years before fisheries could secure a place in the Dáil's busy legislative programme. Introducing the bill, Seán Lemass, Minister for Industry and Commerce, clearly indicated to the Abbey fishermen that their battle had been lost. He emphatically reiterated McGilligan's principle that fishing interests would not be allowed to predominate against the greater interests of power production.

The debate itself was surprisingly low-keyed considering the turbulent times in fisheries during the previous few years. Limerick and Clare TDs put Lemass under a certain amount of pressure, but he was not prepared to concede ground:

> The position is that it is almost impossible to contemplate the adoption of any scheme that does not include acquisition of the various fishing rights of the Abbey

fishermen. Are we to abandon the whole idea of trying to organize and develop the resources of the Shannon fisheries because of the trouble of acquiring the interests of that group? Are we to throw away what may prove to be a very valuable national asset because it happens to be politically inconvenient?

Lemass was adamant: the ESB would be taking over control and management of fisheries on the Shannon and adequate compensation would be paid to interest groups. He argued:

> I think the position of the Abbey net fishermen will be safeguarded in this sense, that they will be awarded fair compensation not merely for the acquisition of their property, but also for the fact that the right to fish with nets is being terminated. There will be no safeguard in the sense that they will be allowed to continue to fish after the operations of the Shannon Fisheries Act, as they fished in the past.

■ Ardnacrusha stands completed at the service of the nation.

In the final stages of the bill, he added:

> The Abbey fishermen … are a community of people who have exercised their right of fishing in the Shannon over a long number of years, and merely to give the present members of that community the present value of their fisheries would be inequitable. Consequently it is proposed to supplement the compensation by another sum to be determined by the arbitrator, a sum which will take into account again the fact that the members of that community are losing their means of livelihood, and that it will not be practicable for them to transfer themselves to some other river and there exercise their skill as fishermen.[18]

The bill quickly passed all its stages by July 1935, when it became law.

But it was not until four years later – after numerous sittings of a public inquiry, court litigations, an amendment to the Shannon Fisheries Act, and considerable personal and family hardship – that the final phases of compensation were paid to the fishermen. It was ten years since they had set out on their campaign to save their industry. They had lost, but in taking on successive governments, the army and the police, they succeeded in creating their own history. The last of the Abbey fishermen passed away a few years ago and the only reminder one sees of them now on the river is the angling cot, special to Limerick and to the memory of those men and their families.

The following is a list of Abbey fishermen, their remarkable nicknames, their addresses and ages, and the amounts of compensation paid to each of them in 1938.[19]

	Name	Nickname	Address	Age	Compensation		
1.	Patrick Clancy	Netter	2 Sheep Street	71	749	6	4
2.	Michael Clancy	---	7 Nicholas Street	63	894	9	0
3.	John Clancy	Gakes	10 Nolans Cottages	61	875	4	8
4.	John Clancy	Sugans	Fish Lane	59	885	0	4
5.	James Clancy	Dick	3 Brown's Lane, Edward St	50	761	14	4
6.	John Clancy	Diddles	1 Glue Yard Lane	31	798	9	4
7.	Gerald Clancy	Riley	Ivy Cottages, King's Island	31	700	9	4
8.	Patrick J. Clancy	Poppy	59 Mungret St	24	585	3	4
9.	Thomas Clancy	Tawdy	1 Robert St	24	637	3	4
10.	James Clancy	Bud/The Yank	3 Brown's Lane, Edward St	23	507	19	0
11.	John Clancy	Mickey pick	3 Brown's Lane, Edward St	22	482	14	2
12.	Michael Clancy Jnr	- - -	1 Glue Yard Lane	27	162	0	0

13.	John Clancy	Cauly	43 Upper Clare St	41	472	0	0
14.	Joseph Clancy	Buckets	Athlunkard St	31	388	0	0
15.	Martin Clancy	Young Martin	Athlunkard St	71	692	0	0
16.	Thomas Hayes	Hackney	4 River Lane	70	797	2	0
17.	Robert Hayes	Napoleon	Campbell's Bow, Broad St	66	935	13	4
18.	Martin Hayes	Rab	Athlunkard St	64	1000	9	0
19.	John Hayes Snr	Bone	6 New Rd, Pennywell	58	346	16	0
20.	Patrick Hayes	Sunlight	24 Broad St	51	626	14	4
21.	Micheal Hayes	Lully	5 Francis Abbey	47	769	12	9
22.	Patrick Hayes	Randy	Island Field	44	769	12	8
23.	Thomas Hayes	Bantrum	Island Field	35	724	2	4
24.	Christy Hayes	Susi/Sonny	2 Watergate	31	635	9	4
25.	Christy Hayes	Rialto	2 Cornmarket Row	31	667	9	4
26.	Micheal Hayes	Starry	21 Mungret St	26	546	12	0
27.	John Hayes Jnr	O.K.	6 New Rd, Pennywell	26	529	12	0
28.	Patrick McNamara	Todsie	3 Sheep St	64	1000	9	0
29.	Peter McNamara	Smuts/Iron Man	Meat Market Lane	60	974	0	2
30.	Patrick J. McNamara	Tons of Money	Sir Harry's Mall	46	769	1	3
31.	Patrick McNamara	Balla	2 Glue Yard Lane	30	513	19	7
32.	Peter McNamara	Boar	3 Fish Lane	37	878	10	1
33.	Robert McNamara	Dutch	2 Creagh Lane	37	817	10	1
34.	Aug McNamara	Rabbit	3 Meat Market Lane	36	787	6	7
35.	James McNamara	Elbows	3 Sheet Street	35	976	2	3
36.	James McNamara	Munchin	O'Halloran's Ln, Thomondgate	33	737	17	11
37.	Joseph McNamara	Beaver	9 Fish Lane	30	448	4	11
38.	Aug McNamara	The Music Man	3 Sheep St	26	648	11	11
39.	Peter McNamara	Peerie	2 Meat Market Lane	25	488	7	7
40.	Joseph McNamara	Young Todsie	3 Sheep St	25	173	0	0
41.	Michael Shanny	Old Mike	Lower Park	73	738	10	7
42.	Patrick Shanny	Vinegar	Island Field	65	819	13	3
43.	John (Sean) Shanny	- - -	Lower Park	55	1040	7	3
44.	Patrick Shanny	Der	Lower Park	53	1058	2	11
45.	John Shanny	Brass Band	Lower Park	54	988	2	11
46.	Michael Shanny	Young Buckshoes	Lower Park	50	873	14	3
47.	Patrick F. Shanny	Feeney	Lower Park	42	788	8	3
48.	Patrick Shanny	Pat the Thatcher	Lower Park	41	823	3	11
49.	James Shanny	Jones	Lower Park	35	787	2	3
50.	John Shanny	Tucker	Lower Park	31	733	9	3
51.	Michael Shanny	Young Mike	Lower Park	27	569	16	3
52.	Michael Shanny	Shirter	Lower Park	26	512	11	11
53.	Patrick Shanny	Woods	Lower Park	24	498	3	3
54.	Joseph Shanny	Young Joe	Lower Park	24	498	3	3
55.	James Shanny	Forty	Lower Park	45	718	0	0

Conclusion

Every institution is the shadow of someone. So is every major project. In the case of the Shannon Scheme that person is Paddy McGilligan. There are those who would argue in favour of Tommy McLaughlin, and one can understand why. But the latter's role was limited when compared to the vast sweep of the Minister's portfolio. It was McGilligan who was charged with the political, financial, commercial, technical, social, cultural and logistical responsibilities for the whole undertaking. The very success of the Scheme and the manner of that success is both proof of and tribute to his performance throughout.

There were many in Leinster House, and outside it, who were not just sceptical about the whole project, but who were openly hostile to it. Some of these represented private enterprise in the electricity industry; others just thought that it was too much of a gamble for the young State, when money was so badly needed for more immediate and pressing needs. There were those too who, while initially being in favour of the Scheme, turned against it because working conditions were so bad there, alleging that the Minister turned out to have feet of clay. But as work progressed and public opinion began to swing in his favour, McGilligan needed to spend less time on his feet in the Dáil and Senate convincing opponents and waverers. His performance there from the very outset was positive, and at times brilliant, speaking on one occasion for seventy-five minutes without notes.

When necessary he was sharp and combative and readily demolished the classic bugbears that the Scheme represented the poisonous virus of nationalization or the cloven hoof of socialism. There were few, if any, at the time who could have equalled his performance in Leinster House. His own background before becoming Minister, in law, in classics, as administrative secretary to Kevin O'Higgins and to the Irish High Commission in London, well qualified him for the task in hand.

McGilligan's papers in the Archives of University College Dublin are very revealing as to the secret of his success. This lay in his intense preparatory work and in his attention to detail. Before embarking on the Shannon Scheme he conducted extensive studies on the industry in other countries. Likewise, before he set up the Electricity Supply Board, he visited North America, examining various organizational models there. The man left little to chance and examined all the options before embarking on a course of action. His ability to choose the right one is evidenced by the international acclaim that the Shannon Scheme received on its completion and the remarkably successful place the ESB has won for itself, not just in our national life during the past seventy-seven years, but on the international stage too.

Of course, Tommy McLaughlin has to be credited with playing a major part in the Shannon Scheme story. A very good friend of McGilligan's from their student days in UCD, it was he who sold the idea to the government. The rest is history. What is not so well known, though, is that McLaughlin advised McGilligan against running for election in 1927 because the latter's health was so poor. Fortunately for the country, the Minister did not take his advice and he remained active in politics for practically another forty years! He was best man at McLaughlin's wedding in Limerick in 1927, right in the middle of construction works, when the latter married Olwen O'Malley, daughter of Joseph O'Malley, a local engineer. It was no great surprise when the ESB was set up in 1927 that McLaughlin was appointed its first managing director. But there were some raised eyebrows four years later when what became a very personalized row broke out between the two men over spending limits at the ESB, resulting in McLaughlin's departure from the Board. He was reappointed by de Valera in 1932, when Fianna Fáil came to power, as there had been considerable public disquiet at his dismissal. It was generally recognized that both men were at fault in the row and their friendship did not suffer any long-term effects.

One of the benefits of anniversaries is that they afford an opportunity not only to revisit events in our history, but also to examine some of the mythology in which they are sometimes shrouded. One often hears it said that the Shannon Scheme was the brainchild of Tommy McLaughlin, and there is no doubting the extensive work he did on it – with Siemens in Berlin and in meetings with the government in Dublin. However, caveats have been entered by some colleagues as to the notion that the project was his brainchild. The late Joe Mac Donald, a Scottish engineer with Shannon Power Development and subsequently with the ESB and Siemens for a time, claimed that the credit for the Scheme lay with Siemens engineers. Mac Donald pointed out in a television interview for the occasion of the fiftieth anniversary that McLaughlin was just out of college and would not have had the experience to mastermind such a project. He also recalled the research on the hydrology of the Shannon extending over three decades of that great engineering pioneer J. Chaloner Smith, who placed all of his findings at Siemens' disposal, a fact acknowledged and highly valued by them.

Whatever the merits or otherwise of Mac Donald's remarks, Siemens came to Ireland with considerable collective experience of similar undertakings in other countries. They had already developed a template for the harnessing of rivers for electricity production. They were also aware that theirs' was not the first plan for Shannon power. Moreover, they knew that the Swiss were involved with a private company to harness the Liffey in order to supply Dublin. So, when they arrived in 1925 there was quite a history to the relevant debate here, and this of itself ratcheted up the pressure on the Germans to succeed. But when things began to go seriously wrong in late 1927, due to local circumstances, and the contractors found themselves falling behind schedule, there was more than money at stake. Siemens requested a six-months extension; McGilligan refused and told them that the circumstances under which the development of the Scheme were undertaken gave it a publicity that was almost worldwide. He urged them to keep on schedule for the prestige of the government and for Siemens. The company replied that for them the Shannon Scheme was more than just a business proposition; their national interest was at stake too. Both sides were in agreement as to what their priorities were, so, in spite of the problems, the contract was finished on time, much to the credit of Siemens. Paddy McGilligan was the hero of the hour and the government was lauded in the international press for its remarkable achievement, against all the odds. The fact that 'one of our own,' Tommy McLaughlin, was involved with Siemens on the engineering side, meant that he too had his own special position on the victors' podium. Considering what the success story did for national self-confidence and morale at the time, and the on-going need for the country to succeed, one can understand if some gave McLaughlin a little more than his well-earned share in the glory. There are those too who would say that with his considerable capacity as a self-publicist he would have claimed it anyway, as he practically did in his Raidió Éireann broadcast in 1938, 'How I Thought of the Shannon Scheme'.

There were others on McGilligan's team who were major players in the drama. Gordon Campbell, Secretary to the Department of Industry and Commerce, was one. His skill and adroitness in negotiations, particularly in the early difficult days of the project, were something to be admired. Professor Rishworth of University College Galway, who acted as Chief Civil Engineer for the government on Shannon Power Development, and who was brought into the team at McGilligan's request, made a major contribution in managing the team's way through difficulties and problems. Incidentally his opposite number on the electrical side, Dr J.E. Sothman, an American engineer, warranted a remark from McLaughlin to McGilligan that his appointment was a grave mistake and that if the Minister went on making appointments like that it would be disastrous for the whole undertaking! One man who interfaced on a daily basis with the contractors on the ground was J.K. Prendergast, the Chief Resident Engineer, in Limerick. He had a most difficult task, with all kinds of problems from the various sites crossing his desk. In a sense he represented an administrative roundabout, as he not just had to cope with the local difficulties but he also had to manage responsibilities upwards. From the copious correspondence one can see that he did a marvellous job administering with fairness and equanimity.

On the minus side, McGilligan and the government fell down badly on living conditions and wages for the workers. Granted, the country was broke and cutbacks were the order of the day. Even the poor pensioners had their money reduced from 10s. to 9s. a week. That was bad enough, and it has gone into the national folklore, but to have thousands of men on starvation wages and hundreds of them living in pigsties and stables could not be justified. Moreover, the local medical services could not cope with the numbers injured and desperately needed extra resources. McGilligan himself admitted that plenty of financial institutions were willing to come in on the project. However, he proved to be just as conservative as the Department of Finance in this matter, and he used his considerable skills to defend the indefensible. In one Dáil debate on the subject he even declared that it was not the function of the government to create jobs! It would be facile to raise a head of moral indignation by imposing a distorting contemporary template on those days. Times were tough then, but nothing could justify what workers had to endure on the project. The Shannon Scheme was a great showcase for the young State and its construction provided a marvellous opportunity for the government to raise standards for workers, even by a modest degree. Sadly they chose not to do so, leaving a legacy of cynicism and bitterness that took years to dispel.

In spite of all the difficulties, the Scheme finished on time, a tribute to the German engineers, administrators, and tradesmen who worked on it. The House of Siemens, as its founder used to call it, did Ireland proud with its magnificent achievement. If development is the growing ability of a people to use and to increase their own resources, then the Germans gave us some purchase and leverage on our future. Their legacy to this country was immense and the story of their sojourn in Ireland has not yet really been told. The project was a milestone in international cooperation and forged new links when both countries badly needed them.

The successful construction of the Scheme was the first step towards industrial and economic independence. It was a beacon of hope and promise. It represented a quantum leap in terms of progress and meant so much for agriculture, the country's main industry. With it came the establishment of the first Irish state venture, the Electricity Supply Board. Numerous other developments have flowed from these bringing the country to where it finds itself today. Ardnacrusha, however, once the supplier of nearly all of the country's power, now, unbelievably, supplies as little as 1 per cent of it – a measure of the distance travelled in seventy-five years. 'The eighth wonder of the world' is now taken for granted. But that is not to detract from the colossal achievement of those who built it. Theirs' is a great story needing to be told on its anniversary and for years to come.

Sources and References

Chapter 1 Did The Germans Really Know?

John Bowman, *De Valera and the Ulster Question, 1917-73* (Oxford 1982).

Conor Brady, *Guardians of the Peace* (Dublin 1974).

Interview with Major General Michael Brennan by author at his home in Killiney, Co. Dublin, 1978.

The Clare Champion, 1918-1926.

Tim Pat Coogan, *Ireland since the Rising* (Dublin 1966).

– *Michael Collins: A Biography* (London 1990).

– *De Valera: Long Fellow, Long Shadow* (London 1993).

Interview with General Michael J. Costello by author at his home in Clontarf, Dublin, 1980.

Ronan Fanning, *The Irish Department of Finance, 1922-58* (Dublin 1978).

David Fitzpatrick, *Politics and Irish Life, 1913-1921: Provincial Experience of War and Revolution* (Dublin 1977).

Tom Garvin, *1922: The Birth of Irish Democracy* (Dublin 1996).

J.A. Gaughan, *Austin Stack* (Kildare 1977).

C. Desmond Greaves, *The Irish Transport and General Workers' Union: The Formative Years* (Dublin 1982).

Michael Hopkinson, *Green Against Green: The Irish Civil War* (Dublin 1988).

The Irish Times, 1918-1926.

Dermot Keogh, *Twentieth-Century Ireland: Nation and State* (Dublin 1994).

The Limerick Chronicle, 1918-1926.

The Limerick Leader, 1918-1926.

Maurice Manning, *The Blueshirts* (Dublin 1987).

McGilligan Papers, University College Dublin Archives Department.

Military Archives, Operation Reports for Clare and Limerick - CW/OPS/2/A, CW/OPS/3/B, CW/OPS/2/D, CW/OPS/2E, CW/OPS/2/F, CW/OPS/2/N, CW/OPS/2/K.

Military Archives, S377/WS 1068 – Statement by Michael Brennan, Column Commander East Clare Flying Column.

National Archives S 7/11/26 – Department of Finance File on Military Protection for the Shannon Scheme containing Major Michael Brennan's report and copious departmental and inter-departmental correspondence.

N.A. S 438/14 – Department of the Taoiseach (President) file containing agenda for meeting of the Executive Council, correspondence and report of General Eoin O'Duffy.

William O'Brien, *Forth the Banners Go* (Dublin 1969).

Leon Ó Broin, *Michael Collins* (Dublin 1980).

– *No Man's Man* (Dublin 1982).

Emmet O'Connor, *Syndicalism in Ireland 1917-1923* (Cork 1988).

Nollaig Ó Gadhra, 'Earnán de Blaghd, 1880-1975', *Éire-Ireland*, 11, no.3 (1976).

Ernie O'Malley, *On Another Man's Wound* (Dublin 1990).

The Singing Flame (Dublin 1979).

John M. Regan, *The Irish-Counter Revolution 1921-1936* (Dublin 1999).

Maryann G. Valuilis, *Portrait of a Revolutionary: General Richard Mulcahy and the Founding of the Irish Free State* (Dublin 1992).

Chapter 2 No Light without Heat

1.   *The Irish Times*, 14 August 1925.
2.   *Irish Independent*, 1 September 1925.
3.   Ordinance Survey Clare 63. 63A.
4.   *Irish Independent*, 11 September 1925.
5.   *The Limerick Leader*, 14 October 1925.
6.   *Irish Independent*, 12 September 1925.
7.   *Irish Independent*, 14 September 1925.
8.   *The Limerick Leader*, 17 October 1925.
9.   *The Voice of Labour*, 5 December 1925.
10. *The Limerick Leader*, 16 September 1925.

11. *Ibid.*, 21 September 1925.
12. *Ibid.*, 16 September 1925.
13. *Ibid.*, 3 October 1925.
14. *Ibid.*, 24 September 1925.
    Ibid., 26 September 1925.
    Ibid., 30 September 1925.
15. *Irish Independent*, 25 September 1925.
16. *Ibid.*, 26 September 1925.
    The Irish Times, 26 September 1925.
17. *The Irish Times*, 29 September 1925.
18. *The Voice of Labour*, October 1925.
19. *Ibid.*, 5 December 1925.
20. *Ibid.*
21. *The Limerick Leader*, 3 October 1925.
22. *The Irish Times*, 2 October 1925.
    The Limerick Leader, 3 October 1925.
    The Irish Times, 8 October 1925.
23. *The Voice of Labour*, 5 December 1925.
24. The Limerick Leader, 6 October 1925.
25. *Ibid.*, 3 October 1925.
26. *Irish Independent*, 9 October 1925.
27. *Irish Independent*, 25 September 1925.
28. *The Limerick Echo*, 6 October 1925.
29. *Irish Independent*, 5 October 1925.
30. *The Limerick Leader*, 5 October 1925.
    The Voice of Labour, 24 October 1925.
31. *Irish Independent*, 7 October 1925.
32. *The Limerick Leader*, 5 October 1925.
    The Irish Times, 3 October 1925.
33. *The Limerick Leader*, 7 October 1925.
34. *The Voice of Labour*, 5 December 1925.
35. *The Limerick Leader*, 14 October 1925.
36. *The Irish Times*, 3 October 1925.
37. *The Limerick Leader*, 10 October 1925.
38. *The Voice of Labour*, 17 October 1925.
39. *Ibid.*, 5 December 1925.
40. *Ibid.*, 11 November 1925.
41. *The Limerick Leader*, 12 October 1925.
42. *The Voice of Labour*, 17 October 1925.
43. *Ibid.*
44. *The Limerick Leader*, 14 October 1925.

45. *Ibid.*
46. *Irish Independent*, 16 October 1925.
47. *The Limerick Leader*, 17 October 1925.
48. *Ibid.*
49. *Irish Independent*, 17 October 1925.
50. *The Irish Times*, 17 October 1925.
51. *Irish Independent*, 19 October 1925.
52. *The Irish Times*, 20 October 2025.
53. *Irish Independent*, 20 October 1925.
54. *The Voice of Labour*, 5 December 1925.
55. *Ibid.*
56. *Ibid.*
57. *Irish Independent*, 26 October 1925.
58. *The Voice of Labour*, 5 December 1925.
59. *Irish Independent*, 27 October 1925.
60. *The Voice of Labour*, 5 December 1925.
61. *Irish Independent*, 31 October 1925.
62. *The Voice of Labour*, 7 November 1925.
63. *Ibid.*
64. *Dáil Debates*, 13.38.
65. *The Manchester Guardian*, 23 October 1925.
66. *The Limerick Leader*, 7 November 1925.
67. *The Voice of Labour*, 21 November 1925.
68. *Irish Independent*, 19 November 1925.
69. *The Limerick Echo*, 10 November 1925.
70. *Irish Independent*, 19 November 1925.
71. *The Limerick Leader*, 28 November 1925.
72. *Ibid*, 23 November 1925.
73. *The Voice of Labour*, 28 November 1925.
74. *Ibid.*
75. *Ibid.*, 12 December 1925.
76. Seanad Debates, 6.34.
77. *Ibid.*, 6.37.
78. *Ibid.*, 6.45.
79. *Ibid.*, 6.46.
80. *The Voice of Labour*, 5 December 1925.
81. *Ibid.*, 12 December 1925.
82. *Irish Independent*, 23 December 1925.
    The Voice of Labour, 16 January 1925.
83. *The Limerick Leader*, 11 January 1925.

84. *Fifty Years of Shannon Power*, ESB Golden Jubilee Publication

Chapter 3 The Land Question

1. The School Manuscripts, Folklore Department, University College Dublin.
2. Senate Report, vol. 5, p.382.
3. McGilligan Papers, P35/30, Memorandum of Inspectors Comyn and McLaughlin, 23 October 1929, Archives Department, University College Dublin.
4. *The Limerick Leader*, 2 September 1929.
5. NA, SS 374, 451, 375, 13558.
6. McGilligan Papers, P35/78.
7. *Ibid.*, P35/30. Gordon Campbell to Secretary Department of Finance, 24 June 1926.
8. *Ibid.*, Seosamh Ua Braonáin to Secretary Department of Industry and Commerce, 26 June 1926.
9. *Ibid.*, Attached copies of correspondence with Finance.
10. *The Limerick Leader*, 4 September 1926.
11. NA, SS 13937.
12. McGilligan Papers, P35/39.
13. NA, SS 657, Secretary Department of Industry and Commerce to Eamon de Valera TD, 21 December 1927.
14. *The Limerick Chronicle*, 14 May 1927.
15. NA, SS 657, op.cit.
16. *The Limerick Leader*, 15 September 1927.
17. *The Limerick Leader*, 29 November 1927.
18. *The Limerick Leader*, 20 August 1928.
19. *The Limerick Leader*, 20 August 1928.
20. *The Limerick Leader*, 2 July 1928.
21. *The Limerick Leader*, 8 June 1929.
22. *The Limerick Leader*, 5 June 1929.
23. *The Limerick Leader*, 15 January 1927.
24. McGilligan Papers, P35/30.
25. NA, SS 13558, John Leydon, Department of Industry and Commerce to Secretary Department of Finance, 12 July 1932.

Chapter 4 As Others See Us

1. NA, SS 13929, Heintze to Rishworth, 27 January 1926.
2. *Ibid.*

3. *Ibid.*, Heintze to Rishworth, 16 February 1926.
4. *Ibid.*, Siemens Bauunion to Rishworth, 21 July 1927.
5. *Ibid.*, Prendergast to Rishworth, 25 June 1926.
6. *Ibid.*
7. *Ibid.*, Resident Engineer to Siemens Schuckertwerke, 29 November 1926.
8. *Ibid.*, Siemens Bauunion to Resident Engineer, 30 November 1926.
9. *Ibid.*, Resident Engineer to Siemens Bauunion, 1 December 1926.
10. *Ibid.*, Dowling to Chief Engineer, 31 January 1927.
11. *Ibid.*, Heintze to Rishworth, 10 February 1927.
12. *Ibid.*, Rishworth to Siemens Schuckertwerke, 14 February 1927.
13. *Ibid.*, file C.6/7, Branch Employment Office, Department of Industry and Commerce, 26 February 1927.
14. *Ibid.*, T.M. Randles to Dowling, 11 July 1927.
15. *Ibid.*, Resident Engineer to Siemens Schuckertwerke, 12 July 1927.
16. *Ibid.*, Siemens Bauunion to Resident Engineer, 14 July 1927.
17. *Ibid.*, Rishworth to Siemens Schuckertwerke, 21 July 1927.
18. NA, SS 13928, List of German Workers.
19. *The Limerick Chronicle*, 19 March 1926.
20. NA, SS 13929, Heintze to Rishworth, 14 April 1927.
21. *Ibid.*, Siemens Bauunion to Chief Engineer, 12 July 1927.
22. Terence Browne, *Ireland: A Social and Cultural History, 1922 to the Present,* (Ithaca NY 1985).
23. NA, SS 13929 Heintze to Rishworth, Enclosures 1 & 2.
24. *Ibid.*, Report of the Trial of Irish Gangs on the Dublin-Kildare 38KV Line, 15 December 1927 – 30 May 1927.
25. *Ibid.*, Randles to Dowling, 20 March 1928.
26. *Ibid.*, Dowling to Resident Engineer, 21 March 1928.
27. *Ibid.*, Randles to Dowling, 8 June 1928.
    Ibid., Dowling to Prendergast, 9 May 1928.
28. *Ibid.*
29. *Ibid.*, J.MacLaughlin to Resident Engineer, 31 July 1929.
30. *Ibid.*, Resident Engineer to Rishworth, Bloomer, MacLaughlin and Sullivan, 6 December 1928.
31. *Ibid.* Goffin to J. McGrath, 15 March 1929.
32. *Ibid.*, Resident Engineer to Siemens Schuckertwerke, 3 June 1929.
33. *Ibid.*
34. *Ibid.*, Secretary, National Ex-Servicemen's League to Resident Engineer, 10 June 1927.
35. *Ibid.*, Rishworth to Bach, 11 July 1929.
36. *Ibid.*, Terence Casey to Chief Engineer, 9 December 1929.
37. *Ibid.*, Prendergast to Siemens Schuckertwerke, 10 December 1929.

38. McGilligan Papers, University Collge Dublin Archives Department, P35/115.

Chapter 5 Living Rough

1. *The Clare Champion*, 8 May 1926
2. *The Limerick Leader*, 5 October 1928
3. *Ibid.*, 7 January 1929
4. *The Engineer*, 16 December 1927
5. *The Clare Champion*, 3 July 1926
6. *Fifty Years of Shannon Power*. ESB Golden Jubilee publication.
7. *The Voice of Labour*, 2 January 1926, p.8.
8. *The Limerick Leader*, 9 May 1928
9. *Ibid.*, 30 June 1926
10. *The Clare Champion*, 8 May 1926
    The Limerick Leader, 10 April 1926
11. *The Clare Champion*, 8 May 1926
12. *Ibid.*, 3 July 1926
13. *Ibid.*
14. *Irish Independent*, 26 June 1926, p.6.
15. *Ibid.*, 29 June 1926, p.7.
16. *The Limerick Leader*, 21 April 1926
17. *The Voice of Labour*, 2 January 1926, p.8.
18. *Ibid.*, 3 July 1926
19. *Ibid.*, 8 May 1926, p.3.
20. *Ibid.*, 12 June 1926, p.2.
21. *The Old Limerick Journal*, December 1980, p.21.
22. *The Voice of Labour*, 24 April 1926, p.4.
    Ibid., 26 June 1926, p.6.
23. *Dáil Debates*, 16 1902-1904
24. *Ibid.*, 16/2017-'8
25. *Ibid.*, 16/2020
26. *Ibid.*, 16/2022
27. *Ibid.*, 16/2024
28. *Ibid.*, 16/2023
29. *Ibid.*, 16/2032/'4
30. *Ibid.*
31. *The Irish Times*, 2 July 1926, p.5.
    The Limerick Leader, 5 July 1926
    The Clare Champion, 17 July 1926

32. *The Irish Times*, 12 July 1926
33. *The Clare Champion*, 17 July 1926
34. *The Limerick Leader*, 12 July 1926
35. *The Clare Champion*, 17 July 1926
36. *Ibid.*, 28 August 1926
37. McGilligan Papers, p.35/30, University College Dublin Archives
38. *The Limerick Leader*, 27 July 1928
39. *Ibid.*, 6 October 1926
40. *The Clare Champion*, 9 October 1926
41. *Ibid.*
42. *Ibid.*
43. *The Limerick Leader*, 8 February 1928
44. *The Clare Champion*, 4 February 1928
45. *Ibid.*, 18 February 1928, p.7.
46. *The Limerick Leader*, 15 October 1928
47. *The Voice of Labour*, 2 January 1926, p.1.

Chapter 6 Trouble by the Mile

1. NA, SS 10467, Secretary, Department of Finance to Secretary, Department of Industry and Commerce, 22 February 1927.
2. *The Limerick Leader*, 24 December 1926.
3. NA, SS10467, *ibid.*, *op.cit.*
4. *Ibid.*, Siemens Bauunion to Prof. Rishworth, 9 September 1927.
5. *Ibid.*, J.Cahill to Bloomer, 21 April 1927.
6. *Ibid.*, Heintze to Rishworth, 13 August 1927.
7. NA, SS 13937.
8. NA, SS 13934, Resident Engineer to Siemens Schuckertwerke, 13 December 1926.
9. *Ibid.*, Siemens Bauunion to Resident Engineer, 17 January 1927.
10. *Ibid.*, Stolberg to Head Office, 11 May 1927.
11. *Ibid.*, Goffin to Resident Engineer, 13 May 1927.
12. *Ibid.*, Rishworth to Resident Engineer, 21 July 1927.
13. *Ibid.*, Stolberg to Head Office, 2 May 1927.
14. *Ibid.*, Keehan to Bloomer, 25 July 1927.
15. *Ibid.*, Heintze to Resident Engineer, 20 July 1927.
16. *Ibid.*, Rishworth to Resident Engineer, 28 July 1927.
17. *Ibid.*, Resident Engineer to Rishworth, 29 July 1927.
    Ibid., Resident Engineer to Rishworth, 17 August 1927.
    Ibid., Rishworth to Resident Engineer, 21 July 1927.

18. *Ibid.*, Department of Industry and Commerce to Boyle, 24 August 1927.
19. *Ibid.*, V.M.McMahon to Resident Engineer, 30 June 1928.
20. *Ibid.*, J.M.Fay, Shannon Power Development, to Resident Engineer, 27 March 1928.
21. *Ibid.*, M.Keehan to Resident Engineer, 30 March 1928.
22. *Ibid.*, V.M.McMahon to Resident Engineer, 30 June 1928.
23. *Ibid.*, M.Keehan to Resident Engineer, 4 February 1929.
24. NA, Limerick Civil Bill Book, 1D/41/94, E1925-H1934.
25. NA, SS 13934 Resident Engineer to Rishworth, 1 November 1926.
26. NA, SS 13937.
27. NA, SS 13934, Siemens Bauunion to Resident Engineer, 20 December 1926.
    Ibid., Siemens Bauunion to Resident Engineer, 20 December 1926 (31/173).
28. *Ibid.*, Resident Engineer to Rishworth, 24 January 1927.
29. *Ibid.*, Dietrich to Head Office, 29 November 1927.
30. *Ibid.*, Rishworth to Resident Engineer, 21 July 1927.
31. *Ibid.*, Taylor to Secretary, Department of Industry and Commerce, 17 January 1928.
32. *Ibid.*, Resident Engineer to Bloomer, 30 January 1928.
33. *Ibid.*, Resident Engineer to Rishworth, 8 June 1928.
34. *Ibid.*, Rishworth to Resident Engineer, 12 June 1928.
35. *Ibid.*, Rishworth to Patrick Clear, 12 June 1928.
36. *Ibid.*, Resident Engineer to Rishworth, 15 October 1929.
37. *The Limerick Leader*, 13 September 1926.

Chapter 7 Counting the Real Cost

1. NA, SS 13927.
2. *Ibid.*, Stolberg to Siemens Bauunion, 30 March 1926.
3. *The Limerick Chronicle*, 25 June 1928.
4. *The Limerick Leader*, 22 April 1926.
5. NA, SS 13927, Report, 2 April 1926.
6. *Ibid.*, Heintze to Rishworth, 25 November 1926.
7. *Ibid.*
8. NA, SS 13928, List of German workers.
9. NA, SS 13927, Heintze to Rishworth, 3 February 1927.
10. *The Cork Examiner*, 19 November 1928.
11. NA, SS 13927, Rishworth to Siemens Bauunion, 14 March 1929.
12. *Ibid.*, Report of M. Keehan, 31 December 1928.
13. *Ibid.*, Shannon Power Development Report on Fatalities for 1926.
14. *The Limerick Chronicle*, 9 November 1926.
    Ibid., 23 November 1926.

15.   *Ibid.*, 14 August 1928.
16.   *Ibid.*, 18 November 1928.
17.   *The Limerick Leader*, 3 May 1926.
18.   *The Limerick Chronicle*, 11 February 1928.
19.   NA, SS 13927, Buckley to Flanagan, 14 December 1926.
20.   *Ibid.*, J.K. Prendergast to Buckley, 17 June 1927.
      Ibid., J.K. Prendergast to McNamara, 20 July 1927.
21.   *Ibid.*, Resident Engineer to J.M.Fay, Shannon Power Development, 12 November 1926.
22.   *Ibid.*
23.   *The Limerick Chronicle*, 2 August 1928.
24.   *The Limerick Leader*, 13 November 1929.
25.   *The Limerick Chronicle*, 21 August 1926.
26.   NA, SS 13928, List of German Workers.
27.   *The Limerick Chronicle*, 24 March 1927.
28.   *Ibid.*, 11 November 1927.
29.   *The Limerick Leader*, 13 October 1928.
      NA, SS 14073, Report into death of Josef Horner.
30.   *The Limerick Leader*, 20 October 1928.
      Ibid., 6 February 1929.
31.   *Ibid.*, 18 May 1929.

Chapter 8   Keeping the Lid on Things

1.   *The Limerick Leader*, 17 July 1929.
2.   *Ibid.*, 14 September 1929.
3.   *Ibid.*, 30 September 1929.
4.   *Ibid.*, 21 October 1929.
5.   *Ibid.*, 14 September 1929.
6.   *Ibid.*, 27 May 1929.
7.   *Ibid.*, 12 July 1926.
8.   *Ibid.*, 12 June 1929.
9.   *The Limerick Chronicle*, 14 May 1927.
10.   *Ibid.*, 7 December 1926.
11.   *The Limerick Leader*, 29 December 1926.
12.   *The Limerick Chronicle*, 24 November 1928.
      The Limerick Leader, 28 November 1928.
13.   *The Limerick Leader*, 14 November 1927.
14.   *Ibid.*, 25 April 1928.

15. *Ibid.*, 21 June 1929.
16. *Ibid.*, 24 April 1928.
17. *Ibid.*, 16 October 1926.
18. *Ibid.*, 14 October 1927.
19. *Ibid.*, 19 January 1929.
20. *Ibid.*
21. *The Limerick Leader*, 12 January 1929.
22. *The Irish Law Times and Solicitors' Journal*, LXIII, No.3246, p.88, Falconer, Dublin.

Chapter 9 Feeling Hard Done By

1.   *The Limerick Leader*, 25 July 1928.
     Ibid., 18 August 1928.
     Ibid., 6 October 1928.
     Ibid., 24 October 1928.
     Ibid., 3 June 1929.
     Ibid., 10 June 1929.
2.   NA, SS 545, Undated Memorandum re. Complaint.
3.   *Ibid.*
4.   *Ibid.*, Minister for Industry and Commerce to John Nolan TD, 12 March 1931.
5.   *Ibid.*, Fr Moloney to President Cosgrove, March 1931.
6.   *Ibid.*, Secretary Department of Industry and Commerce to Secretary, Department of the President, 26 March 1931.
7.   *Ibid.*, Paddy McGilligan to Limerick Diocesan Secretary, 16 March 1931.
8.   *Ibid.*, Limerick Diocesan Secretary to Minister for Industry and Commerce, 11 June 1931.
9.   *Ibid.*, Minister for Industry and Commerce to Limerick Diocesan Secretary, 16 March 1931.
10. *Ibid.*, Resident Engineer to Chief Engineer's office, 4 July 1927.
11. *Ibid.* Resident Engineer to Rishworth, 8 September 1927.
12. *Ibid.*, Department of Industry and Commerce to J.K. Prendergast, 10 August 1927.
13. *Ibid.*, Fr Moloney to Arthur Taylor, 29 July 1927.
14. *Ibid.*, Ciaran Kelly to Department of Industry and Commerce, 11 October 1927.
15. *Ibid.*, Johanna Ross to Department of Industry and Commerce, 4 June 1928.
16. *Ibid.*, Arthur Taylor to Department of Industry and Commerce, 4 March 1931.
17. NA, SS 614, Secretary, Commissioners for Public Works to Department of Finance, 16 September 1928.
18. *Ibid.*, Secretary, Department of Finance to Secretary, Department of Industry and Commerce, 25 June 1928.

19. *Ibid.*, Ms Helen Ryan to Minister for Industry and Commerce, 31 October 1928.
20. *Ibid.*, Secretary, Commissioners for Public Works to Secretary, Department of Industry and Commerce, 5 June 1931.
21. *Ibid.*, Arthur Cox to Secretary, Department of Industry and Commerce, 5 June 1931.
22. *Ibid.*, Rishworth to Secretary, Department of Industry and Commerce.
23. NA, SS 903, Mrs D. Browne to Irish Government Office, 20 May 1930.
24. *Ibid.*, Minute Sheet, 23 October 1931.
25. *Ibid.*, *op. cit.*
26. NA, SS 628, Mrs Davidson to Resident Engineer, 8 June 1927.
27. *Ibid.*, S. Connolly and Co. to Minister for Industry and Commerce, 29 July 1927.
28. *Ibid.*, Arthur Taylor to Secretary, Department of Industry and Commerce, 31 August 1927.
29. NA, SS 618, Arthur Taylor to Secretary, Department of Industry and Commerce, 6 October 1927.
30. *Ibid.*, Secretary, Department of Finance to Secretary, Department of Industry and Commerce, 10 November 1927.
31. *Ibid.*, Minute Sheet, 24 January 1930.
32. *Ibid.*, Prendergast to Rishworth, 21 May 1928.
33. *Ibid.*, Minute Sheet, 24 January 1930.
34. *Ibid.*, Siemens Schuckertwerke to Rishworth, 21 January 1930.
35. *Ibid.*, Minute Sheet (9), 24 January 1930.
36. *Ibid.*, Arthur Taylor to Secretary, Department of Industry and Commerce, 26 February 1930.
37. *Ibid.*, J.J.McElligott to Secretary, Department of Industry and Commerce, 18 July 1930.
38. NA, SS 769, John J. Dundon to P.J.O'Sullivan, Solicitor, 24 May 1929.
39. *Ibid.*, Arthur Taylor to Secretary, Department of Industry and Commerce, 20 July 1929
40. *Ibid.*, Secretary, Department of Finance to Shannon Power Development, 6 August 1929.

Chapter 10 The Battle of the Tailrace

1. Robert Herbert, 'The Lax Weir and Fisheries' Stent of Limerick', *North Munster Antiquarian Journal*, 5 (2 & 3), 1946-47.
2. Government White Paper, 7 March 1924.
3. *The Limerick Chronicle*, 22 January 1927.
4. *The Limerick Leader*, 11 July 1932.
5. *Ibid.*, 9 June 1926.

Ibid., 14 June 1926.

6. *Ibid.*, 22 October 1928.
7. *An Phoblacht*, 16 July 1932.
8. *Dáil Debates*, Vol. 43, 5 July 1932.
9. *The Limerick Leader*, 11 July 1932.
10. *Ibid.*, 13 July 1932.
11. William Lysaght, *The Abbey Fishermen* (Limerick 1964).
12. *An Phoblacht*, 23 July 1932.
13. *The Limerick Leader*, 13 July 1932.
14. *Ibid.*, 18 July 1932.
15. *Ibid.*, 18 July 1932.
16. *Ibid.*, 18 July 1932.
17. *Ibid.*, 22 July 1932.
18. *Dáil Debates*, 1935, Vol. 58.
19. William Lysaght, *The Abbey Fishermen* (Limerick 1964).
20. Interviews by the author with Abbey fishermen in the 1970s and 1980s.

Appendices

Appendix 1

■ Book of Reference No. 1

Townland: Killeely, Co. Limerick

No. on Plan	Description of Property	Owners	Lessees	Name and Address of Occupiers
2	Land	Irish Land Commission (Church Termporalities a/c) R.0.1598	Mary Donnellan wid.	Mary Donnellan (wid.), New Road, Limerick
3 + 4	Land	E.M.H. Seymour & another	Reps. Thos. Donnellan (Mary Donnellan wid.)	Mary Donnellan (wid.), New Road, Limerick
1 + 6	Co. Road	County Council		

Townland: Monabraher, Co. Limerick

No. on Plan	Description of Property	Owners	Lessees	Name and Address of Occupiers
24,25,1, 28,29+30	Land	E.M.H. Seymour & another	Reps. Thos. Donnellan Mary Donnellan (wid.)	Mary Donnellan (wid.), New Road, Killeely, Limerick
2 + 3	Great Southern Railway Line & Embankment	Great Southern Railway Co.		
4, to 18 incl.	Land	E.M.H. Seymour & another	Thos. Hartigan	Thos. Hartigan, Quinsville House, Parteen, Limerick
19 to 23 incl.	Land	E.M.H. Seymour & another	Michael O'Grady	Michael O'Grady, Rossmadda, Ardnacrusha, nr. Limerick
26 + 27	Land	E.M.H. Seymour	Arthur Brennan	Arthur Brennan, Clonconane, nr Limerick
31	Co. Road – (under) Railway – (over)	Limerick Co. Council Great Southern Railway Co.		
32 to 36 incl.	County Road	Limerick Co. Council		

Townland: Ballynanty More, Co. Limerick

No. on Plan	Description of Property	Owners	Lessees	Name and Address
1 + 2	Land	John Hartigan in fee	In fee	John Hartigan, Ballynanty More, nr. Limerick
3	Railway Line	Great Southern Railway Co.		

Townland: Quinspool South, Co. Clare

No. on Plan	Description of Property	Owners	Lessees	Name and Address of Occupiers
1-5 incl. + 8	Land	Patrick Lane	In fee	Patrick Lane, Quinspool, Limerick
5, 6	Public Right of Way (R.O.W.) on foot	Patrick Lane		Patrick Lane, Quinspool, Limerick
7	Water	Patrick Lane		Patrick Lane, Quinspool, Limerick
9 + 11	Land & Osier Plantation	Reps. Jn. McNamara (dec'd)	In fee (registered chargeants for sum of £975)	Reps. Jn. McNamara (dec'd)
10	Water	Anne McNamara (wid.)		Anne McNamara (wid.), c/o John McDonnell, Merchant, Rathkeale
9	County Road	Clare Co. Council		

Townland: Island in Shannon etc., Co. Clare

No. on Plan	Description of Property	Owners	Lessees	Name and Address
1 + 2	Land	Lucy P. Gwynne	In fee	Miss Lucy Penelope Gwynne, Parteenalax, Parteen, Limerick
3	Tideway of R. Shannon	Lax Weir Fishery Co.; John Clancy & Partner		Lax Weir Fishery Co. and John Clancy & Partner
4 + 5	Foreshore adjacent to above	The public exercise a R.O. Passage on foot along foreshore		

Townland: Parteen, Co. Clare

No. on Plan	Description of Property	Owners	Lessees	Name and Address
1 to 6 incl.	Land	Lucy P. Gwynne	In fee	Miss Lucy P. Gwynne (spinster), Parteenlax, Parteen, Limerick
7, 9	Land + garden	H.J. Rycroft	In fee	H.J. Rycroft, Quinsborough House, Parteen, Limerick
10, 11	Land	Gloster Minors; Agents: Messrs Barrington, Limerick	In fee	Gloster Minors (Agents: Messrs Barrington, Barrington st, Limerick)
12	Offices	Gloster Minors; Agents: Messrs Barrington, Limerick	In fee	Gloster Minors (Agents: Messrs Barrington, Barrington st, Limerick)
13	Yard to above	Gloster Minors; Agents: Messrs Barrington, Limerick	In fee	Gloster Minors (Agents: Messrs Barrington, Barrington st, Limerick)

Townland: Parteen, Co. Clare contd.

No. on Plan	Description of Property	Owners	Lessees	Name and Address
14	Land	Gloster Minors; Agents: Messrs Barrington, Limerick	In fee	Gloster Minors (Agents: Messrs Barrington, Barrington st, Limerick)
15	Land	Gloster Minors; Agents: Messrs Barrington, Limerick	In fee	Catherine Davidson (wid.), Parteen, Limerick
16	Land (held with Parteen Nat Schools adjoining)	Very Rev E. Canon Russell PP (as Manager of Parteen Nat. Schools)	in fee	Very Rev. E. Canon Russell PP, Ardnacrusha, nr. Limerick; Ciarán O'Kelly N.T., Parteen, Limerick
4	County Road	Co. Council		

Townland: Ballykeelaun, Co. Clare

No. on Plan	Description of Property	Owners	Lessees	Name and Address
33,34, 35,36	Land	Elizabeth Ryan (wid.)	In fee	Elizabeth Ryan (wid.), Ballinagard, Ballineety P.O., Limerick
30,31,32	Land	Patrick Boyle	In fee	Patrick Boyle, Ballykeelaun, Ardnacrusha, nr. Limerick
26,27,29	Land	Bridget Browne (wid.)	In fee	Bridget Browne (wid.), Ballykeelaun, Ardnacrusha, nr. Limerick
28	Footpath R.O.W	Bridget Browne (wid.)	In fee	Bridget Browne (wid.), Ballykeelaun, Ardnacrusha, nr. Limerick
21,22,23	Land	Thos. Hartigan	In fee	Thos. Hartigan, Quinville House, Parteen, Limerick
24	Dwelling hse in ruins	Thos. Hartigan	In fee	Thos. Hartigan, Quinville House, Parteen, Limerick
25	Yard	Thos. Hartigan	In fee	Thos. Hartigan, Quinville House, Parteen, Limerick
16	Land	R.J.C Maunsell and others	Bridget Curry (wid.)	Bridget Browne (wid.), Ballykeelaun, Ardnacrusha, nr. Limerick
18, 19	Land	R.J.C Maunsell and others	Andrew Dundas	Andrew Dundas, Ballykeelaun, Ardnacrusha, Limerick
20 not completed on sheet	Land	Bridget Curry (wid.)	In fee	Bridget Curry (wid.), Ballykeelaun, Ardnacrusha, nr. Limerick
17	Land	R.J.C Maunsell and others	Martin Hamilton	Martin Hamilton, Ballykeelaun, Ardnacrusha, nr. Limerick

Townland: Ballykeelaun, Co. Clare contd.

No. on Plan	Description of Property	Owners	Lessees	Name and Address
23 to 27 incl.	County Road	Co. Council		
6,7,8,9, 10,11,13	Land	Rep. of Michael Curry (dec'd), Bridget Curry (wid.)		Bridget Curry (wid.), Ballykeelaun, Ardnacrusha, nr. Limerick
3	Land	R.J.C Maunsell & others and I.L.C	Andrew Dundas	Andrew Dundas, Ballykeelaun, Ardnacrusha, Limerick
4	Dwelling hse, offices and ??	R.J.C Maunsell & others and I.L.C	Andrew Dundas	Andrew Dundas, Ballykeelaun, Ardnacrusha, Limerick

Townland: Castlebank, Co. Clare

No. on Plan	Description of Property	Owners	Lessees	Name and Address
1,2,3	Land	Patrick McCormack	In fee	Patrick McCormack, Castlebank House, Parteen, nr. Limerick
4	Well	Patrick McCormack	In fee	Patrick McCormack, Castlebank House, Parteen, nr. Limerick
5	Mass Path R.O.W.	Patrick McCormack	In fee	Patrick McCormack, Castlebank House, Parteen, nr. Limerick
9, 10	Land	Patrick McCormack	In fee	Patrick McCormack, Castlebank House, Parteen, nr. Limerick
11 not completed	Land	Robert Holmes	In fee	Robert Holmes, Rose House, Athlunkard, Limerick
5,6,7,8	Land	Robert Holmes and ILC		Robert Holmes, Rose House, Athlunkard, Limerick

Townland: Parkroe, Co. Clare

No. on Plan	Description of Property	Owners	Lessees	Name and Address
1	Land	Capt. P.B. McAdam + ILC	Patrick McMahon Snr.	Patrick McMahon, Blackwater, Parteen, Limerick

Townland: Ballykeelaun, Co. Clare

No. on Plan	Description of Property	Owners	Lessees	Name and Address
6,7,8,9, 10, 11,13	Land	Rep. of Michael Curry (dec'd), Bridget Curry (wid.) and ILC	None	Same as owners, Ballykeelaun, Ardnacrusha, nr. Limerick

Townland: Ballykeelaun, Co. Clare contd.

No. on Plan	Description of Property	Owners	Lessees	Name and Address
3	Land	R.J.C Maunsell and others and ILC		Andrew Dundas (also lessee), Ballykeelaun, Ardnacrusha, Limerick
4	Dwelling house, offices and yard	R.J.C Maunsell and others and ILC		Andrew Dundas (also lessee), Ballykeelaun, Ardnacrusha, Limerick

Townland: Castlebank, Co. Clare

No. on Plan	Description of Property	Owners	Lessees	Name and Address
9, 10	Information supplied	Robert Holmes and ILC	vested	Robert Holmes, Rose House, Athlunkard, Limerick
5,6,7,8	Land	Robert Holmes and ILC		Robert Holmes, Rose House, Athlunkard, Limerick

Townland: Parkroe, Co. Clare

No. on Plan	Description of Property	Owners	Lessees	Name and Address
1	Land	Capt. P.B. McAdam + ILC	Patrick McMahon Snr	Patrick McMahon, Blackwater, Parteen, Limerick

Townland: Killeely, Co. Limerick

No. on Plan	Description of Property	Owners	Lessees	Name and Address
1	Land	ILC (Church Temporalities a/c) R.O. 1598	Mary Donnellan (wid.)	Mary Donnellan
2	County Road	Limerick Co. Council		
3	Land	E.M.H. Seymour & J.N. Seymour	Reprs. Thomas Donnellan (Mary Donnellan)	Mary Donnellan

■ Book of Reference No. 2

Townland: Parkroe, Co. Clare   Barony: Bunratty Lower

No. on Plan	Description of Property	Owners	Lessees	Name and Address
1	Land	Capt. P.B. McAdam	Patrick McMahon	Patrick McMahon

Townland: Ballykeelaun Co. Clare — Barony: Bunratty Lower

No. on Plan	Description of Property	Owners	Lessees	Name and Address
3	Land	R.J.C.Maunsell & M.N. Maunsell (spinster)	Andrew Dundas	Andrew Dundas
4	Dwelling House, Offices and Yard	R.J.C.Maunsell & M.N. Maunsell (spinster)	Andrew Dundas	Andrew Dundas
5	County Road	Clare Co. Council		
6,7,8,9, 10,11,13	Land	Rep. of Michael Curry (dec'd), Bridget Curry (widow)		Bridget Curry (widow)
15 & part of 14	Land (Labourers' plots)	Clare Co. Council		Joseph Keegan & Thomas Wolf
12, 16, & part of 14	Land	R.J.C.Maunsell & M.N. Maunsell (spinster)	Bridget Curry (widow)	Bridget Curry (widow)
17	Land	R.J.C.Maunsell & M.N. Maunsell (spinster)	Martin Hamilton	Martin Hamilton
18 + 19	Land	R.J.C.Maunsell & M.N. Maunsell (spinster)	Andrew Dundas	Andrew Dundas
20	Land	Bridget Curry (widow)		Bridget Curry (widow)
21,22,23 24 25	Land Dwelling hse in ruins Yard	Thomas Hartigan		Thomas Hartigan
26,27,29 28	Land Footpath	Bridget Browne (widow)		Bridget Browne (widow)
30,31,32	Land	Patrick Boyle		Patrick Boyle
33,34, 35,36	Land	Elizabeth Ryan (widow)		Elizabeth Ryan (widow)

Townland: Castlebank Co. Clare — Barony: Bunratty Lower

No. on Plan	Description of Property	Owners	Lessees	Name and Address
5,6,7,8	Land	Robert Holmes		Robert Holmes
9, 10	Land	Patrick McCormack		Patrick McCormack
11	Land	Robert Holmes		Robert Holmes
12,13,14	Land	Patrick McCormack		Patrick McCormack
15	Footpath	Patrick McCormack		Patrick McCormack

Townland: Garraun Co. Clare — Barony: Bunratty Lower

No. on Plan	Description of Property	Owners	Lessees	Name and Address
1,2	Land	Patrick Clear		Patrick Clear
3	County Road	Clare Co. Council		Clare Co. Council

Townland: Garraun Co. Clare contd.   Barony: Bunratty Lower

No. on Plan	Description of Property	Owners	Lessees	Name and Address
4	Avenue	Timothy Crowe		Dr James Brennan
5 + 6	Land	Timothy Crowe		Timothy Crowe
7	Gate Lodge	Timothy Crowe		Dr James Brennan
8	Avenue	Timothy Crowe		Dr James Brennan
9	Orchard Garden (part of)	Timothy Crowe		Dr James Brennan
10	Land	Timothy Crowe		Timothy Crowe

Townland: Parteen Co. Clare   Barony: Bunratty Lower

No. on Plan	Description of Property	Owners	Lessees	Name and Address
1	Land	Gloster Minor Agents: J.B. Barrington, Limerick		Gloster Minors
2	Land	Very Rev. E. Canon Russell PP (Mgr of Parteen Nat. Schools)		Ciarán O'Kelly N.T.
3	Land	Gloster Minor Agents: J.B. Barrington, Limerick		Gloster Minors
4	Road	Clare Co. Council, Patrick McCormack & Gloster Minors		Clare Co. Council, Patrick McCormack & Gloster Minors
5 6 + 11	Garden Lands	H.J. Rycroft		H.J. Rycroft
7	Yard	Gloster Minor Agents: J.B. Barrington, Limerick		Gloster Minors
8	Offices	Gloster Minor Agents: J.B. Barrington, Limerick		Gloster Minors
9 + 10	Land	Gloster Minor Agents: J.B. Barrington, Limerick		Gloster Minors
12,13,14, 15,16 &17	Land	Lucy P. Gwynne (spinster)		Lucy P. Gwynne (spinster)
18	County Road	Clare Co. Council		Clare Co. Council

Townland: River Shannon, Co. Clare

No. on Plan	Description of Property	Owners	Lessees	Name and Address
1+ 2	Foreshore of R. Shannon	Saorstát Éireann		Saorstát Éireann

Townland: Quinspool South, Co. Clare   Barony: Bunratty Lower

No. on Plan	Description of Property	Owners	Lessees	Name and Address
1 + 2	Land + osier plantation	Reprs. John McNamara (dec'd), Anne McNamara (widow)		Reprs. John McNamara (dec'd), Anne McNamara (widow) c/o John McDonnell, Merchant, Rathkeale
3	Stream Part of 1 + 4	Reprs. John McNamara (dec'd), Anne McNamara (widow) & Patrick Lane		Reprs. John McNamara (dec'd), Anne McNamara (widow) c/o John McDonnell & Patrick Lane
4,6,7,8	Land	Patrick Lane		Patrick Lane
5	Footpath	Patrick Lane		Patrick Lane
9	County Road	Clare Co. Council		Clare Co. Council

Townland: Monabraher, Co. Limerick   Barony: North Liberties

No. on Plan	Description of Property	Owners	Lessees	Name and Address
1 + 2	Land	E.M.H. Seymour & J.N. Seymour	Reprs. Thomas Donnellan (Mary Donnellan)	Mary Donnellan
3	Land	E.M.H. Seymour & J.N. Seymour	Arthur Brennan	Arthur Brennan
4,5,6,7,8	Land	E.M.H. Seymour & J.N. Seymour	Michael O'Grady	Michael O'Grady
9 + 16	Railway	Great Southern Railway Co.		Great Southern Railway Co.
10	County Road	Limerick Co. Council		Limerick Co. Council
11	County Road	Limerick Co. Council		Limerick Co. Council
12,13, 14,15	Land	E.M.H. Seymour & J.N. Seymour	Thomas Hartigan	Thomas Hartigan

Townland: Killeely   County Borough of Limerick

No. on Plan	Description of Property	Owners	Lessees	Name and Address
1	Land	ILC (Church Temporalities a/c R.O. 1598	Mary Donnellan (widow)	Mary Donnellan (widow)
2	County Road	Limerick Corporation		Limerick Corporation
3	Land	E.M.H. Seymour & J.N. Seymour	Reprs. Thomas Donnellan (Mary Donnellan)	Mary Donnellan (widow)

■ Book of Reference No. 3

Townland: Knockbrack Lower, Co. Clare

No. on Plan	Description of Property	Owners	Lessees	Name and Address
1	Land	Patrick Crowe		Patrick Crowe
2,3,4,5,6	Land	Arthur Stritch		Arthur Stritch
7	Out office	Arthur Stritch		Arthur Stritch
8,9,10	Land	Peter Mulqueen		Peter Mulqueen
11	Land	Nora Sheedy (spinster)		Nora Sheedy (spinster)
12	Land	Denis Toomey		Denis Toomey
13,14, 15,16	Land	Jas. Jos. Guinane		Jas. Jos. Guinane
17	Cart Passage	Nora Sheedy (spinster)		Nora Sheedy (spinster)
18	? County Road	Clare Co. Council		Clare Co. Council
19	County Road	Clare Co. Council		Clare Co. Council

Townland: Cloonlara, Co. Clare

No. on Plan	Description of Property	Owners	Lessees	Name and Address
1,2,3	Land	Edward N. Murphy		Edward N. Murphy
4	Land	Clare Co. Council		James Walsh
5	Dwelling house & Yard	Clare Co. Council		James Walsh
6 to 16 incl.	Land	Patrick Stritch		Patrick Stritch
18	Footpassage & right to take water from well	The inhabitants of the district		The inhabitants of the district
19,20, 21,22	Land	Reprs. Maurice O'Sullivan (dec'd), Michael O'Sullivan		Reprs. Maurice O'Sullivan (dec'd), Michael O'Sullivan
23,24	Land	Arthur Stritch		Arthur Stritch
25	Land & disused quarry	Arthur Stritch		Arthur Stritch
26	Land	Reprs. Maurice O'Sullivan (dec'd), Michael O'Sullivan		Michael O'Sullivan
27	Dwelling house & garden	Reprs. Maurice O'Sullivan (dec'd), Michael O'Sullivan		Michael O'Sullivan
28	Creamery Buildings & Yard	Reprs. Maurice O'Sullivan (dec'd), Michael O'Sullivan		The Condensed Milk Co. of Ireland, Ltd.
29	County Road	Clare Co. Council		Clare Co. Council

Townland: Cloonlara, Co. Clare contd.

No. on Plan	Description of Property	Owners	Lessees	Name and Address
30	County Road	Clare Co. Council		Clare Co. Council
31	Dwelling House (in ruins)	John Moloney		John Moloney
32	Land	Bridget Hannan (widow)		Bridget Hannan (widow)
33	Cart Passage	Reps. R.S. Walsh (dec'd), Col. H.M. Butler, Mrs Alice Walsh		Reps. R.S. Walsh (dec'd), Col. H.M. Butler, Mrs Alice Walsh
34	County Road	Clare Co. Council		Clare Co. Council
35	County Road	Clare Co. Council		Clare Co. Council
36	County Road	Clare Co. Council		Clare Co. Council
37	Telephone Line	Min. for Post and Telegraphs		Min. for Post and Telegraphs

Townland: Newtown, Co. Clare

No. on Plan	Description of Property	Owners	Lessees	Name and Address
1	Land	Reps. R.S. Walsh (dec'd), Col. H.M. Butler, Mrs Alice Walsh		Reps. R.S. Walsh (dec'd), Col. H.M. Butler, Mrs Alice Walsh
2	Land	Reps. R.S. Walsh (dec'd), Col. H.M. Butler, Mrs Alice Walsh		Reps. R.S. Walsh (dec'd), Col. H.M. Butler, Mrs Alice Walsh
3	Quarry (disused)	Reps. R.S. Walsh (dec'd), Col. H.M. Butler, Mrs Alice Walsh		Reps. R.S. Walsh (dec'd), Col. H.M. Butler, Mrs Alice Walsh
4	Land	Reps. R.S. Walsh (dec'd), Col. H.M. Butler, Mrs Alice Walsh		Reps. R.S. Walsh (dec'd), Col. H.M. Butler, Mrs Alice Walsh
5	½ County Road	Clare Co. Council		Clare Co. Council
6	Land & Avenue	Reps. R.S. Walsh (dec'd), Col. H.M. Butler, Mrs Alice Walsh		Reps. R.S. Walsh (dec'd), Col. H.M. Butler, Mrs Alice Walsh
7	Land	Reps. R.S. Walsh (dec'd), Col. H.M. Butler, Mrs Alice Walsh		Reps. R.S. Walsh (dec'd), Col. H.M. Butler, Mrs Alice Walsh
8	Telephone Line	Min. for Post and Telegraphs		Min. for Post and Telegraphs

Townland: Oakfield, Co. Clare

No. on Plan	Description of Property	Owners	Lessees	Name and Address
1,3	Land	John O'Shaughnessy Jnr.		John O'Shaughnessy Jnr.

Townland: Oakfield, Co. Clare contd.

No. on Plan	Description of Property	Owners	Lessees	Name and Address
2	Land	Reps. R.S. Walsh (dec'd), Col. H.M. Butler, Mrs Alice Walsh		Col. H.M. Butler, Mrs Alice Walsh
4	½ County Road	Clare Co. Council		Clare Co. Council
5	Telephone Line	Min. for Post and Telegraphs		Min. for Post and Telegraphs

Townland: Ruanard, Co. Clare

No. on Plan	Description of Property	Owners	Lessees	Name and Address
1,2,3, 4,5,6	Land	Reps. R.S. Walsh (dec'd), Col. H.M. Butler, Ms F.M. Walsh		Reps. R.S. Walsh (dec'd), Col. H.M. Butler, Ms F.M. Walsh
7	Land	Reps. R.S. Walsh (dec'd), Col. H.M. Butler, Ms F.M. Walsh		John Hynes
8	Land	Reps. R.S. Walsh (dec'd), Col. H.M. Butler, Ms F.M. Walsh		Reps. R.S. Walsh (dec'd), Col. H.M. Butler, Ms F.M. Walsh
9	Avenue	Reps. R.S. Walsh (dec'd), Col. H.M. Butler, Ms F.M. Walsh		Reps. R.S. Walsh (dec'd), Col. H.M. Butler, Ms F.M. Walsh
10	Gate Lodge	Reps. R.S. Walsh (dec'd), Col. H.M. Butler, Ms F.M. Walsh		Reps. R.S. Walsh (dec'd), Col. H.M. Butler, Ms F.M. Walsh
11	Land & Woodland	Reps. R.S. Walsh (dec'd), Col. H.M. Butler, Ms F.M. Walsh		Reps. R.S. Walsh (dec'd), Col. H.M. Butler, Ms F.M. Walsh
12	Dwelling house, out-offices & yard	Reps. R.S. Walsh (dec'd), Col. H.M. Butler, Ms F.M. Walsh		John Hynes
13	Cart Passage	Reps. R.S. Walsh (dec'd), Col. H.M. Butler, Ms F.M. Walsh		Reps. R.S. Walsh (dec'd), Col. H.M. Butler, Ms F.M. Walsh
14	½ County Road	Clare Co. Council		Clare Co. Council
15	Land & Footpath	Reps. R.S. Walsh (dec'd), Col. H.M. Butler, Ms F.M. Walsh		Reps. R.S. Walsh (dec'd), Col. H.M. Butler, Ms F.M. Walsh
16	Land	Reps. R.S. Walsh (dec'd), Col. H.M. Butler, Ms F.M. Walsh		James Hynes
17	Dwelling house & garden	Col. H.M. Butler, Mrs Alice Walsh		James Hynes
18	Land	Reps. R.S. Walsh (dec'd), Col. H.M. Butler, Ms F.M. Walsh		Col. H.M. Butler, Ms F.M. Walsh

■ Book of Reference No. 4

Townland: Ballykeelaun, Co. Clare   Barony: Bunratty Lower

No. on Plan	Description of Property	Owners	Lessees	Name and Address
15	Land	Clare Co. Council		Reps. of John Keegan (dec'd), Nora Keegan (widow), Michael Keegan
37	Land	Capt. R.J.C.Maunsell & M.N. Maunsell (spinster)		Reps. of Martin Hamilton (dec'd), John Hamilton
38	Dwelling house, out-offices & yard	Clare Co. Council		John Prenderville
39	Land	Clare Co. Council		John Prenderville
40	Land	Capt. R.J.C.Maunsell		Michael O'Grady
41	Dwelling house, out-offices, yard & garden	Clare Co. Council		Reps. of John Keegan (dec'd), Nora Keegan (widow), Michael Keegan
42	County Road	Clare Co. Council		Clare Co. Council
43	½ County Road	Clare Co. Council		Clare Co. Council

Townland: Parkroe, Co. Clare   Barony: Bunratty Lower

No. on Plan	Description of Property	Owners	Lessees	Name and Address
2 + 4	Land	Capt. P.B. McAdam		Reps. of Elizabeth Keane, Mary Keane (widow)
3	½ County Road	Clare Co. Council		Clare Co. Council

Townland: Reana-Brone, Co. Clare   Barony: Tulla Lower

No. on Plan	Description of Property	Owners	Lessees	Name and Address
1	Land	Dudley Persse Joynt.		John Hodgins
2	Avenue	Dudley Persse Joynt.		John Hodgins
3	Gate Lodge	Dudley Persse Joynt.		John Hodgins
4	½ County Road	Clare Co. Council		Clare Co. Council

Townland: Blackwater, Co. Clare   Barony: Bunratty Lower

No. on Plan	Description of Property	Owners	Lessees	Name and Address
1,2,3,4	Land	Capt. P.B. McAdam		Capt. P.B. McAdam
5	½ Weir	Capt. P.B. McAdam		Capt. P.B. McAdam

Townland: Ballyglass, Co. Clare   Barony: Tulla Lower

No. on Plan	Description of Property	Owners	Lessees	Name and Address
1 to 10 incl.	Land	William J. Holmes		William J. Holmes
11	Land	Mary E. Maunsell		Michael O'Reilly
12	Dwelling house, out-offices & yard	Mary E. Maunsell		Michael O'Reilly
13, 14	County Road	Clare Co. Council		Clare Co. Council
15	½ County Road	Clare Co. Council		Clare Co. Council

Townland: Rosmadda West, Co. Clare   Barony: Bunratty Lower

No. on Plan	Description of Property	Owners	Lessees	Name and Address
1,2,3, 4,5	Land	Capt. R.J.C.Maunsell & M.N. Maunsell (spinster)		Michael O'Grady
6	Dwelling house, out-offices & yard	Capt. R.J.C.Maunsell & M.N. Maunsell (spinster)		Michael O'Grady
7	Land	Capt. R.J.C.Maunsell	Michael O'Grady	Michael O'Grady
8	Land	Capt. R.J.C.Maunsell	Michael O'Grady	Michael O'Grady
9	Dwelling house	Capt. R.J.C.Maunsell		James Hayes
10	Land	Capt. R.J.C.Maunsell		Stanley Jas. Johnstone
11	Orchard	Capt. R.J.C.Maunsell		Stanley Jas. Johnstone
12	Yard & garden	Capt. R.J.C.Maunsell		Stanley Jas. Johnstone
13	Dwelling house & out-offices	Capt. R.J.C.Maunsell		Stanley Jas. Johnstone
14	Land	Capt. R.J.C.Maunsell		Stanley Jas. Johnstone
15	Mill (in ruins)	Capt. R.J.C.Maunsell		Stanley Jas. Johnstone
16	R.O.W.	Capt. R.J.C.Maunsell		Michael O'Grady
17	Land	William J. Holmes		William J. Holmes
18, 19	Land	Capt. R.J.C.Maunsell	Courtenay Croker	Courtenay Croker
20	Cottage	Capt. R.J.C.Maunsell & M.N. Maunsell (spinster)	Courtenay Croker	Courtenay Croker
21	½ weir	Capt. R.J.C.Maunsell & M.N. Maunsell (spinster)	Courtenay Croker	Courtenay Croker
22	Hydraulic Ram	Capt. R.J.C.Maunsell & M.N. Maunsell (spinster)	Courtenay Croker	Courtenay Croker
23	½ County Road	Clare Co. Council		Clare Co. Council
24	County Road	Clare Co. Council		Clare Co. Council

Townland: Rosmadda West, Co. Clare contd.   **Barony: Bunratty Lower**

No. on Plan	Description of Property	Owners	Lessees	Name and Address
25	Telephone Line	Min. for Post and Telegraphs		Min. for Post and Telegraphs
26	Sluice Gates	Capt. R.J.C.Maunsell & M.N. Maunsell (spinster)	Courtenay Croker	Courtenay Croker
27	Mill Race	Stanley J Johnstone		Stanley J Johnstone, Michael O'Grady, Courtenay Croker & William J Holmes

■ Book of Reference No. 5

Townland: Ballyglass, Co. Clare   **Barony: Bunratty Lower**

No. on Plan	Description of Property	Owners	Lessees	Name and Address
16 to 25 incl.	Land	Capt. P.B. McAdam (agent H.B. Alton)		Dr Joseph Humpreys
26	Dwelling house, out-offices and yard	Capt. P.B. McAdam (agent H.B. Alton)		Min. for Posts & Telegraphs
27	County Road	Clare Co. Council		Clare Co. Council
28	Telephone Line	Min. for Posts & Telegraphs		Min. for Posts & Telegraphs

Townland: Mount Catherine, Co. Clare   **Barony: Tulla Lower**

No. on Plan	Description of Property	Owners	Lessees	Name and Address
1,2,3	Land	Capt. R.J.C. Maunsell		John J. de Courcy
4	Dwelling house, out-offices & yard	Clare Co. Council		Thomas Whelan
5,6	Land	Clare Co. Council		Thomas Whelan
7	County Road	Clare Co. Council		Clare Co. Council

Townland: Newtown, Co. Clare   **Barony: Tulla Lower**

No. on Plan	Description of Property	Owners	Lessees	Name and Address
9	Land & Avenue	Reps R.S Walsh (dec'd), Mrs Alice Walsh (wid.) & Col. H.M. Butler		Mrs Alice Walsh (wid.) & Col. H.M. Butler
10,11, 12,13	Land	Reps R.S Walsh (dec'd), Mrs Alice Walsh (wid.) & Col. H.M. Butler		Mrs Alice Walsh (wid.) & Col. H.M. Butler

Townland: Newtown, Co. Clare contd. Barony: Tulla Lower

No. on Plan	Description of Property	Owners	Lessees	Name and Address
14	Dwelling house, out-offices and garden	Reps R.S Walsh (dec'd), Mrs Alice Walsh (wid.) & Col. H.M. Butler		James Guinane
15	Land	Reps R.S Walsh (dec'd), Mrs Alice Walsh (wid.) & Col. H.M. Butler		James Guinane
16	Dwelling house, out-offices and yard	Clare Co. Council		James O'Connor
17,18	Land	Clare Co. Council		James O'Connor
19	$^1/_2$ County Road	Clare Co. Council		Clare Co. Council

Townland: Oakfield, Co. Clare Barony: Tulla Lower

No. on Plan	Description of Property	Owners	Lessees	Name and Address
6	Land	Timothy Ryan		Timothy Ryan
7	Land	Clare Co. Council		John Fahy
8	Dwelling house, out-offices and yard	Clare Co. Council		John Fahy
9	Land & Avenue	Timothy Ryan		Timothy Ryan
10	Land	Timothy Ryan		Timothy Ryan
11	Land	John O'Shaughnessy Jnr.		John O'Shaughnessy Jnr.
12	Land & Avenue	John O'Shaughnessy Jnr.		John O'Shaughnessy Jnr.
13	Out-office	John O'Shaughnessy Jnr.		John O'Shaughnessy Jnr.
14	$^1/_2$ County Road	Clare Co. Council		Clare Co. Council
15	Telephone Line	Min. for Posts & Telegraphs		Min. for Posts & Telegraphs

Townland: Knockbrack, Co. Clare Barony: Tulla Lower

No. on Plan	Description of Property	Owners	Lessees	Name and Address
20	County Road	Clare Co. Council		Clare Co. Council
21,22,23 24 & 25	Land	James Jos. Guinane		James Jos. Guinane
26, 27	Land	Peter Mulqueen		Peter Mulqueen

Townland: Monaskeha, Co. Clare   Barony: Tulla Lower

No. on Plan	Description of Property	Owners	Lessees	Name and Address
1	Land	James Jos. Guinane		James Jos. Guinane
2	Footpath	Inhabitants of the District		Inhabitants of the District
3	Land	John O'Dwyer		John O'Dwyer
4	Out office	John O'Dwyer		John O'Dwyer
5	County Road	Clare Co. Council		Clare Co. Council
6, 7	Land	Arthur Stritch		Arthur Stritch
8	Footpath	Inhabitants of the District		Inhabitants of the District

■ Book of Reference No. 6

Townland: Ardclooney, Co. Clare   Barony: Tulla Lower

No. on Plan	Description of Property	Owners	Lessees	Name and Address
1 to 19 incl.	Land	Reps. of Annie Scanlon (dec'd), Rev. Patrick Scanlon		Reps. of Annie Scanlon (dec'd), Rev. Patrick Scanlon
20	Dwelling house (in ruins)	Reps. of Annie Scanlon (dec'd), Rev. Patrick Scanlon		Reps. of Annie Scanlon (dec'd), Rev. Patrick Scanlon
21	Yard & out-offices (in ruins)	Reps. of Annie Scanlon (dec'd), Rev. Patrick Scanlon		Reps. of Annie Scanlon (dec'd), Rev. Patrick Scanlon
22	Roadway	Reps. of Annie Scanlon (dec'd), Rev. Patrick Scanlon		Reps. of Annie Scanlon (dec'd), Rev. Patrick Scanlon
23, 24	Land	Reps of Roger Moloney (dec'd), Roger Moloney Jnr		Reps of Roger Moloney (dec'd), Roger Moloney Jnr
25	Roadway	Reps of Roger Moloney (dec'd), Roger Moloney Jnr		Reps of Roger Moloney (dec'd), Roger Moloney Jnr, also R.O.W. for Reps. of Annie Scanlon (dec'd)

Townland: Ross, Co. Clare   Barony: Tulla Lower

No. on Plan	Description of Property	Owners	Lessees	Name and Address
1	Land	Ambrose Hall	James O'Malley	James O'Malley

Townland: O'Briensbridge, Co. Clare　　　　　Barony: Tulla Lower

No. on Plan	Description of Property	Owners	Lessees	Name and Address
1 to 13 incl.	Land	Lieut. Col. Loftus Bryan, Mrs Marian Vaughan, Mrs Louisa Plunkett Little-Hamilton, Doctor Daniel H Foley	Maria Ryan	Maria Ryan
14, 15	Ruined Buildings	Lieut. Col. Loftus Bryan, Mrs Marian Vaughan, Mrs Louisa Plunkett Little-Hamilton, Doctor Daniel H Foley	Maria Ryan	Maria Ryan

Townland: Birdhill, Co. Tipperary　　　　　Barony: Kilnamanagh Upper

No. on Plan	Description of Property	Owners	Lessees	Name and Address
1, 3	Land	Jeremiah O'Sullivan		Jeremiah O'Sullivan
2	Engine Shed (with pump well & oil engine)	Jeremiah O'Sullivan		Jeremiah O'Sullivan
4, 5	Land	William Coffee Jnr		William Coffee Jnr
6	Land	Jeremiah O'Sullivan		William J. Riordan

■ Book of Reference No. 7

Townland: Monaskeha, Co. Clare　　　　　Barony: Tulla Lower

No. on Plan	Description of Property	Owners	Lessees	Name and Address
9	Land	John O'Dwyer		John O'Dwyer
10?11,12	Land	James Campbell		James Campbell
13	County Road	Clare Co. Council		Clare Co. Council
14	Land	Arthur Stritch		Arthur Stritch
15?, 16, 17,18	Land	Michael Stritch		Michael Stritch
19	Land	Commissioners of Public Works (CPW)		David Madden
20	Land	CPW		Simon McEvoy

Townland: Bartleystown, Co. Clare Barony: Tulla Lower

No. on Plan	Description of Property	Owners	Lessees	Name and Address
1	Land	James Campbell		James Campbell
2	Land	Michael Kennedy		Michael Kennedy
3	Land	John Ringrose		John Ringrose

Townland: Aughboy, Co. Clare Barony: Tulla Lower

No. on Plan	Description of Property	Owners	Lessees	Name and Address
1 to 14 incl.	Land	Synan Campbell		Synan Campbell
15	Footpath	The Inhabitants of the District		The Inhabitants of the District
16	Building (in ruins)	Synan Campbell		Synan Campbell
17	Land	John Bourke		John Bourke
18, 19	Land	James Corbett		James Corbett
20	½ County Road	Clare Co. Council		Clare Co. Council
21, 22	County Road	Clare Co. Council		Clare Co. Council
23	Land	CPW		David Madden
24	Land	CPW		Patrick Ryan

Townland: Drummeen, Co. Clare Barony: Tulla Lower

No. on Plan	Description of Property	Owners	Lessees	Name and Address
1, 2	Land	P. Ryan		P. Ryan

Townland: Kildoorus, Co. Clare Barony: Tulla Lower

No. on Plan	Description of Property	Owners	Lessees	Name and Address
1	Land	John Bourke		John Bourke
???, 5, 6, 7, 10, ???, 12, 13	Land	Synan Campbell		Synan Campbell
3	Lime Kiln	Synan Campbell		Synan Campbell
8	County Road	Clare Co. Council		Clare Co. Council
9	½ County Road	Clare Co. Council		Clare Co. Council
???,15,16, 17,18,???, 20,21,22, 23,???,25, 23,???,25, 26,27,28	Land	John Gorman		John Gorman

Townland: Kildoorus, Co. Clare contd.   Barony: Tulla Lower

No. on Plan	Description of Property	Owners	Lessees	Name and Address
34, 35, 36, 37, 38,???, 40,41, 42, 43,44,???, 46,47,48,49	Land	Andrew Conlon		Andrew Conlon
???, 31, 32,33	Land	Bridget McMahon		Bridget McMahon
50	House (in ruins)	Andrew Conlon		Andrew Conlon
51	Roadway	Andrew Conlon		Andrew Conlon
52	Land	CPW		Lizzie Collins

Townland: Ardnataggle, Co. Clare   Barony: Tulla Lower

No. on Plan	Description of Property	Owners	Lessees	Name and Address
1	Land	John Daly		John Daly
???, 3,to 12 incl.	Land	Lawrence Cusack		Lawrence Cusack
13	County Road	Clare Co. Council		Clare Co. Council
14,15,16, 17 & 18	Land	Michael Cusack		Michael Cusack
19	Land	Spaight Estate		Michael Cusack
20	County Road	Clare Co. Council		Clare Co. Council
21,22, 23,24	Land	Spaight Estate		John Hastings
25	County Road	Clare Co. Council		Clare Co. Council
26	Footpath	The Inhabitants of the District		The Inhabitants of the District
27	Land	Spaight Estate		Michael Cooney
28,29	Land	Spaight Estate		Annie Hastings
30	Land	Irish Land Trust (Soldiers & Sailors)		William McCarthy
31	Land	CPW		Lizzie Collins
32	Roadway	Lawrence Cusack		Lawrence Cusack

Townland: O'Briensbridge, Co. Clare   Barony: Tulla Lower

No. on Plan	Description of Property	Owners	Lessees	Name and Address
16	Right of Way (ROW)	Thos R. Ryan		Thos. R. Ryan
17	Land	Clare Co. Council		Bridget McLaughlin

Townland: O'Briensbridge, Co. Clare contd. Barony: Tulla Lower

No. on Plan	Description of Property	Owners	Lessees	Name and Address
18 to 26 incl.	Land	Thos. R. Ryan		Thos. R. Ryan
27, 28	Land	Denis Minihan		Denis Minihan
29	House (vacant)	Denis Minihan		Denis Minihan
30	County Road	Clare Co. Council		Clare Co. Council
??, 33,34, 35,36	Land	Patrick Moloney		Patrick Moloney
32	Passage to bog	The Inhabitants of the District		The Inhabitants of the District
37	Garden	Going Estate		Matthew Burke
38	Garden	John Ryan		John Ryan
41 to 46 incl.	Land	Michael McNamara		Michael McNamara
40	Dwelling house, out-offices and yard	Michael McNamara		Michael McNamara
47	Land	John Ryan		John Ryan
48	Passage to bog	The Inhabitants of the District		The Inhabitants of the District
50 to 55 incl.	Land	Michael Ryan		Michael Ryan
56	Dwelling house & out- offices	Michael Ryan		Michael Ryan
57	Garden	Going Estate		John Woulfe
??, 59, 60,61	Land	Ellen Ahern		Ellen Ahern
62	Garden	John Ryan		John Ryan
??, 64, 65	Land	Thos. G Ryan		Thos. G Ryan
66	Passage to bog	The Inhabitants of the District		The Inhabitants of the District
67,68,69	Garden	Thos. G Ryan		Thos. G Ryan
70,71,72	Land	James King		James King
73	ROW	John Quinlivan		John Quinlivan
74,75	Land	John Quinlivan		John Quinlivan
76,77	Land	James Crotty		James Crotty
78 to 82 incl.	Land	James Ryan		James Ryan
83	County Road	Clare Co. Council		Clare Co. Council
84	Land	McCraith Estate		Daniel Fitzgerald
85	Land	McCraith Estate		Michael McNamara

Townland: O'Briensbridge, Co. Clare contd.   Barony: Tulla Lower

No. on Plan	Description of Property	Owners	Lessees	Name and Address
86	Land	McCraith Estate		Delia McKeogh
87	Land	Clare Co. Council		Michael Corbett
88	Dwelling house & out-offices	Clare Co. Council		Michael Corbett
89	Land	Clare Co. Council		Michael Ryan
90	Dwelling house & out-offices	Clare Co. Council		Michael Ryan
91	Land	Joseph Ahern		Joseph Ahern
92	Land	Davis Estate		Delia McKeogh
93	ROW	Michl Cooney, John Hastings		John Ryan
94	Land	John Ryan		John Ryan
95	Garden	Going Estate	Thomas G. Ryan	J. J. Frizelle
96	Garden	Going Estate	Thomas G. Ryan	Mary Roche
97	Garden	Going Estate	Thomas G. Ryan	Michael Spillane
98	Garden	Going Estate	Thomas G. Ryan	Patrick Healy
99	Garden	Going Estate		James Ryan
100	Dwelling house, out-offices and yard	Going Estate		James Ryan
101	County Road	Clare Co. Council		Clare Co. Council
102	Land	Lt. Col. Loftus Bryan, Mrs Marie Vaughan, Mrs Louisa Plunkett Little-Hamilton, Dr Daniel H. Foley	Marie Ryan	Marie Ryan

■ Book of Reference No. 8

Townland: Ardclooney, Co. Clare   Barony: Tulla Lower

No. on Plan	Description of Property	Owners	Lessees	Name and Address
26,27,28	Land	Irish Land Commission (ILC)		Roger Maloney
29	Right of Way (ROW)	Reps. Annie Scanlon (dec'd), Rev P Scanlon & Timothy Murnane		Reps. Annie Scanlon (dec'd), Rev P Scanlon & Timothy Murnane
30 to 38 incl.	Land	Reps. Annie Scanlon (dec'd), Rev P Scanlon & Timothy Murnane		Timothy Murnane

Townland: Ardclooney, Co. Clare contd.　　Barony: Tulla Lower

No. on Plan	Description of Property	Owners	Lessees	Name and Address
39	Dwelling house, out-offices & yard	Reps. Annie Scanlon (dec'd), Rev P Scanlon & Timothy Murnane		Timothy Murnane
40	Land	Reps. Annie Scanlon (dec'd), Rev P Scanlon & Timothy Murnane		Peter Kelly
41 to 60 incl.	Land	ILC		Laurance Flynn
61	Dwelling house, out-offices & land	ILC		Laurance Flynn
62	Out-office	ILC		Laurance Flynn
63 to 84 incl.	Land	ILC		Maurice O'Brien
85	Land	Commissioners of Public Works (CPW)		CPW
86	Land	CPW		Timothy Murnane
87	Land	CPW		John Crowe

Townland: Bannow Islands, Co. Tipperary　　Barony: Owney & Arra

No. on Plan	Description of Property	Owners	Lessees	Name and Address
1,2	Land	CPW		Miss Lizzie Collins

Townland: Illaunanadderha Island, Co. Clare　　Barony: Tulla Lower

No. on Plan	Description of Property	Owners	Lessees	Name and Address
1	Land	ILC		Reps. of P Farrell
2	Land	CPW		John Crowe

Townland: Cloonfadda, Co. Clare　　Barony: Tulla Lower

No. on Plan	Description of Property	Owners	Lessees	Name and Address
1 to 14 incl.	Land	ILC		Reps. of P. Farrell
15 to 19 incl.	Land	ILC		Miss Nora Martin
20	Land	ILC		J.D. O'Callaghan
21	ROW	Reps. of P. Farrell, J O'Callaghan, Norah Martin		Reps. of P. Farrell, J. O'Callaghan, Norah Martin (Spr.)

Townland: Cloonfadda, Co. Clare contd. Barony: Tulla Lower

No. on Plan	Description of Property	Owners	Lessees	Name and Address
22	Land	ILC		J O'Callaghan
23 to 27 incl.	Land	ILC		Reps. of P Farrell
28	Dwelling house, out-offices and yard	ILC		Reps. of P Farrell
29 to 40 incl.	Land	ILC		Reps. of P Farrell
41	Land	ILC		Michael O'Brien
42,43,44	Land	ILC		John Barry
45	Land	ILC		John Crowe
46,47	Land	ILC		John Barry
48	ROW	Michael O'Brien, Denis Crowe & John Crowe		Michael O'Brien, Denis Crowe & John Crowe
49 to 53 incl.	Land	ILC		Michael O'Brien
54	ROW	Denis Crowe & John Crowe		Denis Crowe & John Crowe
55,56, 57,58	Dwelling house, out-office & yard	ILC		Denis Crowe
59	Land	ILC		Denis Crowe
60 to 67 incl.	Land	ILC		Denis Crowe
68	ROW	John Crowe		John Crowe
69,70,71	Land	ILC		Annie Mulcahy
72	ROW	John Crowe		John Crowe
73	ROW	John Crowe		John Crowe
74,75	Land	ILC		Miss Rebecca O'Callaghan
76	ROW	John Crowe		John Crowe
77,78, 79,80	Land	ILC		Miss Rebecca O'Callaghan
81	ROW	John Crowe		John Crowe
82, 84	Land	ILC		Annie Mulcahy
83	ROW	ILC		Annie Mulcahy
85	Land	CPW		John Magee
86,87,88	Land	CPW		John Crowe

Townland: Moys, Co. Clare **Barony: Tulla Lower**

No. on Plan	Description of Property	Owners	Lessees	Name and Address
1	Land	CPW		John Magee
2	Land	CPW		John Crowe
3	Dwelling house, out-offices & Land	CPW		John Crowe
4	Land	Dr William Courtenay		Dr William Courtenay
5	ROW	John Crowe		John Crowe
6,7,8	Land	Dr William Courtenay		Dr William Courtenay

Townland: Birdhill Co. Tipperary **Barony: Kilnamanagh Upper**

No. on Plan	Description of Property	Owners	Lessees	Name and Address
6	Land	Jeremiah O'Sullivan		William Reardon
7,8,9,10	Land	Jeremiah O'Sullivan		Jeremiah O'Sullivan
11	Land	Jeremiah O'Sullivan		William Reardon
12	Out-offices, Dwelling house & yard	Jeremiah O'Sullivan		William Reardon
13	Land	Jeremiah O'Sullivan		William Reardon
14 to 22 incl.	Land	ILC		Mrs Mary Foley
23	Land	ILC		Michael Murnane
24	Land	ILC		Mrs Kate Gleeson
25	Land	ILC		Patrick Hassett
26	Land	ILC		Terence O'Brien
27	Land	ILC		Isaac Buckley
28	Land	ILC		Patrick Gleeson
29	Land	ILC		William Hassett
30	Land	ILC		Patrick Daly
31	Land	ILC		John Ryan
32	Land	ILC		Patrick Grady
33	Land	ILC		John McMahon
34	Land	ILC		Mrs Margaret Carroll
35	Land	ILC		Denis Gleeson
36	Land	ILC		Thomas Buckley
37,38	Land	ILC		Robert Thorne
39	Land	ILC		Michael Kenna

Townland: Birdhill, Co. Tipperary contd.   Barony: Kilnamanagh Upper

No. on Plan	Description of Property	Owners	Lessees	Name and Address
40	Land	ILC		Patrick Ryan
41,42	Land	CPW		John Crowe
43	Land	ILC		Timothy Coffey
44	Land	ILC		John McCormack
45	Land	ILC		Patrick Healy
46	Land	ILC		Patrick Gleeson
47	Land	ILC		Edward Gleeson
48	Land	ILC		Edward Gough
49	½ river	Kilmastulla Drainage Board		Kilmastulla Drainage Board

Townland: Pollagh (E.D. Birdhill), Co. Tipperary   Barony: Owney & Arra

No. on Plan	Description of Property	Owners	Lessees	Name and Address
1	Land	ILC		Patrick Gleeson
2	ROW	The Inhabitants of the District		The Inhabitants of the District
3	Land	Martin Ryan		Martin Ryan
4	Land	ILC		Mrs Jane Ryan
5	Land	ILC		Patrick Browne
6	Land	ILC		Phillip Ahern, Cragg
7	Land	ILC		Michael Murnane
8	Land	ILC		Phillip Ahern
9	Land	ILC		Phillip Ahern, Cragg
10	Land	ILC		Patrick Gleeson
11,12	Land	ILC		Patrick Browne
13,14	Land	ILC		Edmond Bourke
15	Land	ILC		Michael Murnane
16	Land	ILC		John Foley
17,18	Land	ILC		Thomas Gleeson
19	Land	ILC		Edmond Bourke
20	Land	ILC		Jeremiah Keogh
21	Land	ILC		Honora Burke
22	Land	ILC		Jas. Tuohy
23	Land	ILC		Michael O'Hara
24	Land	ILC		Matthew Buckley
25	Land	ILC		Daniel Sammon

Townland: Pollagh (E.D. Birdhill), Co. Tipperary contd.   Barony: Owney & Arra

No. on Plan	Description of Property	Owners	Lessees	Name and Address
26	Land	ILC		Mrs Sarah Buckley
27	Land	ILC		Denis Sammon
28	Land	ILC		Jeremiah Keogh
29	Land	ILC		John Gleeson
30	Land	ILC		Jas. Meaghar
31	Land	ILC		Maria Mullins (spr.)
32	Land	ILC		Patrick Carroll
33,34	Land	ILC		Michael Hanley
35	Land	ILC		John Foley

Townland: Coolnadornory, Co. Tipperary   Barony: Owney & Arra

No. on Plan	Description of Property	Owners	Lessees	Name and Address
1 to 10 incl.	Land	ILC		Isaac Floyd
11	ROW	Catherine Hayes		Catherine Hayes
12,13	Land	ILC		Catherine Hayes
14	½ river	Kilmastulla Drainage Board		Kilmastulla Drainage Board

Townland: Garrynatineel, Co. Tipperary   Barony: Owney & Arra

No. on Plan	Description of Property	Owners	Lessees	Name and Address
1	½ river	Kilmastulla Drainage Board		Kilmastulla Drainage Board
2	Land	CPW		John Crowe
3,4,5	Land	Samuel Whipp	D.L. Cooke	D.L. Cooke
6 to 15 incl.	Land	ILC		D.L. Cooke
16	Dwelling house, out-offices	ILC		D.L. Cooke
17	Out-offices	ILC		D.L. Cooke
18	Hay Barn	ILC		D.L. Cooke
19	½ river	ILC		D.L. Cooke
20	Eel weir (disused)	CPW		Anthony Mackey
21	Land	ILC		Patrick Ryan

■ Book of Reference No. 9

Townland: Parteen, Co. Clare   Barony: Bunratty Lower

No. on Plan	Description of Property	Owners	Lessees	Name and Address
19,20, 21,22	Land	Gloster Minors (J.B. Barrington Agents)		Gloster Minors
23	Dwelling house	Gloster Minors (J.B. Barrington Agents)		Gloster Minors
24	Out-offices & yard	Gloster Minors (J.B. Barrington Agents)		Gloster Minors
25	Garden	Gloster Minors (J.B. Barrington Agents)		Mrs Catherine Davidson
26	Yard & Entrance	Gloster Minors (J.B. Barrington Agents)		Mrs Catherine Davidson
27	Dwelling house	Gloster Minors (J.B. Barrington Agents)		Mrs Catherine Davidson
28	House	Limerick Co. Council		Limerick Co. Council
29	Land	Rev E. Canon Russell PP (Mgr Parteen National Schools)		Ciaran O'Kelly N.T.
30	Yard & out-offices	Rev E. Canon Russell PP (Mgr Parteen National Schools)		Ciaran O'Kelly N.T.
31	Land	Rev E. Canon Russell PP (Mgr Parteen National Schools)		Ciaran O'Kelly N.T.
32	Schools	Rev E. Canon Russell PP (Mgr Parteen National Schools)		Rev E. Canon Russell PP (Mgr Parteen National Schools)
33	Dwelling house	Rev E. Canon Russell PP (Mgr Parteen National Schools)		Ciaran O'Kelly N.T.
34	Dwelling house	Rev E. Canon Russell PP (Mgr Parteen National Schools)		Mrs George Ross
35	Land (Garden & Playground)	Rev E. Canon Russell PP (Mgr Parteen National Schools)		Rev E. Canon Russell PP (Mgr Parteen National Schools)
36	Out-offices	Rev E. Canon Russell PP (Mgr Parteen National Schools)		Rev E. Canon Russell PP (Mgr Parteen National Schools)
37,38, 39,40	Land	Lucy P. Gwynne (spinster)		Lucy P. Gwynne (spinster)
41	Shed	Lucy P. Gwynne (spinster)		Lucy P. Gwynne (spinster)
42	Dwelling House	Lucy P. Gwynne (spinster)		Lucy P. Gwynne (spinster)
43	Lodge	Lucy P. Gwynne (spinster)		Lucy P. Gwynne (spinster)
44	Out-offices	Lucy P. Gwynne (spinster)		Lucy P. Gwynne (spinster)

Townland: Parteen, Co. Clare contd.   **Barony: Bunratty Lower**

No. on Plan	Description of Property	Owners	Lessees	Name and Address
45	St Patrick's Church	Church Representative Body of Ireland		Church Representative Body, Church of Ireland represented by Diocese of Limerick
46	Grounds	Church Representative Body of Ireland		Church Representative Body, Church of Ireland represented by Diocese of Limerick
47	ROW	Public		Public
48	ROW	Public		Public
49	ROW	Owners of Weir		Owners of Weir
50	Concrete Chamber for pump	Lucy P. Gwynne (spinster)		Lucy P. Gwynne (spinster)
51	Road	Clare Co. Council		Clare Co. Council

Townland: Island in R. Shannon

No. on Plan	Description of Property	Owners	Lessees	Name and Address
1	Land	Lucy P Gwynne (spinster)		Lucy P Gwynne (spinster)

Townland: Weir etc. in R. Shannon

No. on Plan	Description of Property	Owners	Lessees	Name and Address
1	Buildings	The Lax Weir Fishing Co. Ltd		The Lax Weir Fishing Co. Ltd
2	Weir	The Lax Weir Fishing Co. Ltd		The Lax Weir Fishing Co. Ltd

Townland: Garraun, Co. Clare   **Barony: Bunratty Lower**

No. on Plan	Description of Property	Owners	Lessees	Name and Address
11	Land & garden	Patrick Clear		Patrick Clear
12	Dwelling house	Patrick Clear		Patrick Clear
13	Out-offices and yard	Patrick Clear		Patrick Clear
14	Road	Clare Co. Council		Clare Co. Council

Townland: Newtown (E.D. Kiltenanlea), Co. Clare   **Barony: Tulla Lower**

No. on Plan	Description of Property	Owners	Lessees	Name and Address
20 to 29 incl.	Land	Reprs. R.S. Walsh (dec'd), Col. H.M Butler & Mrs Alice Walsh		Reprs. R.S. Walsh (dec'd), Col. H.M Butler & Mrs Alice Walsh

Townland: Mountcatherine, Co. Clare   Barony: Tulla Lower

No. on Plan	Description of Property	Owners	Lessees	Name and Address
8, 9	Land	Capt. & Mrs Maunsell		John Wixted
10, 11	Land	Capt. & Mrs Maunsell		Patrick McDonnell
12,13	Land	Capt. R.J.C Maunsell		J.J. De Courcey
14	Land	Capt. & Mrs Maunsell		Patrick Murnane

SUPPLEMENTAL TO REFERENCE FOR ABOVE PREVIOUSLY FURNISHED

Townland: Parteen

No. on Plan	Description of Property	Owners	Lessees	Name and Address
1, 2	Land	Gloster Minors (Agents: J.B. Barrington, Limerick)	In fee	Gloster Minors
3	Land, flower/ kitchen gardens & orchard	Gloster Minors (Agents: J.B. Barrington, Limerick)	In fee	Gloster Minors
4	Land (lawn & pleasure ground)	Gloster Minors (Agents: J.B. Barrington, Limerick)	In fee	Gloster Minors
5	Dwelling house	Gloster Minors (Agents: J.B. Barrington, Limerick)	In fee	Gloster Minors
6	Offices	Gloster Minors (Agents: J.B. Barrington, Limerick)	In fee	Gloster Minors
7	Yard	Gloster Minors (Agents: J.B. Barrington, Limerick)	In fee	Gloster Minors
8	Land	Gloster Minors (Agents: J.B. Barrington, Limerick)	In fee	Gloster Minors
9	Flower garden	Gloster Minors (Agents: J.B. Barrington, Limerick)		Ms Catherine Davidson, Parteen, nr. Limerick
10	Yard & Entrance	Gloster Minors (Agents: J.B. Barrington, Limerick)		Ms Catherine Davidson, Parteen, nr. Limerick
11	Dwelling house and shop (licensed premises)	Gloster Minors (Agents: J.B. Barrington, Limerick)		Ms Catherine Davidson, Parteen, nr. Limerick
12	Building (Parteen Dispensary)	Limerick Co. Council		Limerick Co. Council
13	Land	Very Rev. E. Canon Russell PP (Mgr. Parteen Nat'l Schools)		Ciaran O'Kelly N.T., Parteen, nr. Limerick
14	Yard	Very Rev. E. Canon Russell PP (Mgr. Parteen Nat'l Schools)		Ciaran O'Kelly N.T., Parteen, nr. Limerick

Townland: Parteen contd.

No. on Plan	Description of Property	Owners	Lessees	Name and Address
15, 16	Out-offices	Very Rev. E. Canon Russell PP (Mgr. Parteen Nat'l Schools)		Ciaran O'Kelly N.T., Parteen, nr. Limerick
17	Land (garden & entrance to schools)	Very Rev. E. Canon Russell PP (Mgr. Parteen Nat'l Schools)		Ciaran O'Kelly N.T., Parteen, nr. Limerick, Mrs George Ross, Parteen, nr. Limerick
18	Buildings (schools & residences for teachers)	Very Rev. E. Canon Russell PP (Mgr. Parteen Nat'l Schools)		(18b)Ciaran O'Kelly N.T., Parteen, nr. Limerick & (18c)Mrs George Ross, Parteen, nr. Limerick
19	Land (garden)	Very Rev. E. Canon Russell PP (Mgr. Parteen Nat'l Schools)		Ciaran O'Kelly N.T., Parteen, nr. Limerick, Mrs George Ross, Parteen, nr. Limerick
20	Land (Playground)	Very Rev. E. Canon Russell PP (Mgr. Parteen Nat'l Schools)		Ciaran O'Kelly N.T., Parteen, nr. Limerick, Mrs George Ross, Parteen, nr. Limerick
21	Office (a privy)	Very Rev. E. Canon Russell PP (Mgr. Parteen Nat'l Schools)		Ciaran O'Kelly N.T., Parteen, nr. Limerick, Mrs George Ross, Parteen, nr. Limerick
47	½ of stream	Very Rev. E. Canon Russell PP (Mgr. Parteen Nat'l Schools)		Ciaran O'Kelly N.T., Parteen, nr. Limerick, Mrs George Ross, Parteen, nr. Limerick
22-28	Land (lawn, pleasure ground; flower & kitchen gardens, orchard,	Lucy P. Gwynne (spinster), Parteenalax, nr. Limerick		Lucy P. Gwynne (spinster)
30,30, 32,37	Land	Lucy P. Gwynne (spinster), Parteenalax, nr. Limerick		Lucy P. Gwynne (spinster)
33	Dwelling house	Lucy P. Gwynne (spinster), Parteenalax, nr. Limerick		Lucy P. Gwynne (spinster)
34	Lodge	Lucy P. Gwynne (spinster), Parteenalax, nr. Limerick		Lucy P. Gwynne (spinster)
35, 36	Out-offices	Lucy P. Gwynne (spinster), Parteenalax, nr. Limerick	In fee	Lucy P. Gwynne (spinster)
29	Shed (for pump)	Lucy P. Gwynne (spinster), Parteenalax, nr. Limerick	In fee	Lucy P. Gwynne (spinster)
38	Concrete chamber	Lucy P. Gwynne (spinster), Parteenalax, nr. Limerick	In fee	Lucy P. Gwynne (spinster)
44	Walks & Paths	Lucy P. Gwynne (spinster), Parteenalax, nr. Limerick	In fee	Lucy P. Gwynne (spinster)
48	½ of stream	Lucy P. Gwynne (spinster), Parteenalax, nr. Limerick	In fee	Lucy P. Gwynne (spinster)

Townland: Parteen contd.

No. on Plan	Description of Property	Owners	Lessees	Name and Address
39,40,42	ROW Passages	The Public	NB. These ROWs are	The Public
41	ROW Footpath	Owners of Weir	on the lands of Mrs	Owners of the Weir
43	ROW Passage	Owners of St Patrick's Church	L.P. Gwynne	Owners of St Patrick's Church
45	Road	Clare Co. Council		The Public

SUPPLEMENTAL TO REFERENCE FOR ABOVE PREVIOUSLY FURNISHED

Townland: Island in R. Shannon

No. on Plan	Description of Property	Owners	Lessees	Name and Address
1	Land (under timber)	Lucy P. Gwynne (spinster)		Lucy P. Gwynne (spinster)
2	Land	Lucy P. Gwynne (spinster)		Lucy P. Gwynne (spinster)

SUPPLEMENTAL TO REFERENCE FOR ABOVE PREVIOUSLY FURNISHED

Townland: Weir etc in R. Shannon

No. on Plan	Description of Property	Owners	Lessees	Name and Address
1	Buildings	The Lax Weir Fishing Co.Ltd (Corbally). Secretary: W.O Tottingham, 33 Monument st, London EC3		The Lax Weir Fishing Co.Ltd (Corbally). Secretary: W.O. Tottingham, 33 Monument St, London EC3
2	Weir	The Lax Weir Fishing Co.Ltd (Corbally). Secretary: W.O Tottingham, 33 Monument st, London EC3		The Lax Weir Fishing Co.Ltd (Corbally). Secretary: W.O. Tottingham, 33 Monument St, London EC3
3	Water			

Townland: Garraun

No. on Plan	Description of Property	Owners	Lessees	Name and Address
1, 2	Land	Patrick Clear, Garraun, Parteen, nr. Limerick	In fee	Patrick Clear
3	Garden	Patrick Clear, Garraun, Parteen, nr. Limerick	In fee	Patrick Clear
4	Dwelling house & shop	Patrick Clear, Garraun, Parteen, nr. Limerick	In fee	Patrick Clear
5, 6	Out-offices	Patrick Clear, Garraun, Parteen, nr. Limerick	In fee	Patrick Clear

Townland: Garraun contd.

No. on Plan	Description of Property	Owners	Lessees	Name and Address
7	Yard	Patrick Clear, Garraun, Parteen, nr. Limerick	In fee	Patrick Clear
8	Stream	Patrick Clear, Garraun, Parteen, nr. Limerick	In fee	Patrick Clear
9	½ of stream	Clare Co. Council		The Public
10	Road	Clare Co. Council		The Public

■ Book of Reference No. 10

Townland: Portumna Demesne & Rogers Island, Co. Galway   Barony: Longford

No. on Plan	Description of Property	Owners	Lessees	Name and Address
1	Dumping ground	Portumna Town Committee		Portumna Town Committee
1	Island	Portumna Town Committee		Portumna Town Committee
1	Island	Portumna Town Committee		Portumna Town Committee
'A'	Small Island	Portumna Town Committee		Portumna Town Committee
'B'	Small Island	Portumna Town Committee		Portumna Town Committee

Townland: Portumna, Co. Galway   Barony: Longford

No. on Plan	Description of Property	Owners	Lessees	Name and Address
1	Land	John Taylor		John Taylor
2	Land	John Taylor		John Taylor
3	Land	Miss Meade		Miss Meade
4	Land	Miss Meade		Miss Meade
5	Land	Jos. Coone		Jos. Coone
6	Land	Portumna Town Committee		Portumna Town Committee
7	Land	Jos. Doherty		Jos. Doherty
8	Land	John J. Costello		John J. Costello
9	Land	Patrick Felee		Patrick Felee
10	Land	John J. Martin		John J. Martin
11	ROW	Portumna Town Committee, John Duffy, Michael Derwan, Patrick Power, Bgt. Walsh (wid.), Jos. Madden		
12	Land	Patrick Head	John Duffy	John Duffy

Townland: Portumna, Co. Galway contd. Barony: Longford

No. on Plan	Description of Property	Owners	Lessees	Name and Address
13	Land	Michael Derwan		Michael Derwan
14	Land	Patrick Power		Patrick Power
15	Land	Patrick Power		Patrick Power
16	Land	Bgt. Walsh (wid.)		Bgt. Walsh (wid.)
17	Land	Jos. Madden		Jos. Madden

Townland: Fairyhill, Co. Galway Barony: Longford

No. on Plan	Description of Property	Owners	Lessees	Name and Address
1	Land	Board of Works (BOW)	Jas. Columbia	Jas. Columbia
2	Land	Walter J Joyce		Walter J Joyce
3	Land	Walter J Joyce		Walter J Joyce
4	Land	Walter J Joyce		Walter J Joyce
5	County Road	Galway Co. Council		Galway Co. Council
6	Office	BOW	Jas. Columbia	Jas. Columbia
7	Wharf	BOW		Public
8	Storehouse	BOW	Capt. O. Waller	Capt. O. Waller
9	Storehouse	BOW	Grand Canal Co.	Grand Canal Co.
10	Land	BOW	Jas. Columbia	Jas. Columbia
11	Storehouse	BOW		Jas. Bannatyne
12	Storehouse	BOW		Mrs J.A. O'Brien
13	Yard & Wharf	BOW		BOW
14	ROW	BOW, Thos. Shaughnessy, Ml. P. Tuohy, Fergus Molloy, John Dillon, John Molloy, Mary Mitchell (wid.), Ellen Burke (wid.)		
15	Storehouse	BOW		Jas. Bannatyne
16	Land	BOW	Fergus Molloy	Fergus Molloy
17	Land	Ellen Burke (wid.)		Ellen Burke (wid.)
18	Land	BOW	Thos. Shaughnessy	Thos. Shaughnessy
19	Land	BOW	Thos. Shaughnessy	Thos. Shaughnessy
20	Land	BOW	Thos. Shaughnessy	Thos. Shaughnessy
21	Land	Michael P. Tuohy		Michael P. Tuohy
22	Land	Michael P. Tuohy		Michael P. Tuohy
23	Land	John Molloy		John Molloy

Townland: Fairyhill, Co. Galway contd.   Barony: Longford

No. on Plan	Description of Property	Owners	Lessees	Name and Address
24	Land	John Dillon		John Dillon
25	Land	Mary Mitchell (wid.)		Mary Mitchell (wid.)
26	Land	John Molloy		John Molloy
27	Land	Mary Mitchell (wid.)		Mary Mitchell (wid.)
28	Land	Ellen Burke (wid.)		Ellen Burke (wid.)

Townland: Gortaha, Co. Galway   Barony: Longford

No. on Plan	Description of Property	Owners	Lessees	Name and Address
1	Land	John G. Mullany		John G. Mullany
2	Land	Mrs Mgt. Flannery		Mrs Mgt. Flannery
3	Land	John Molloy		John Molloy
4	Land	(Reps. Maria Maher) (Fergus Molloy)		(Reps. Maria Maher) (Fergus Molloy)
5	Land	Thos. Madden		Thos. Madden
6	Land	Thos. Cooney		Thos. Cooney
7	Land	Thos. Cooney		Thos. Cooney
8	Land	Patrick Kelly		Patrick Kelly
9	Land	Patrick Kelly		Patrick Kelly
10	Land	Thos. Keane		Thos. Keane
11	Land	Thos. Keane		Thos. Keane
12	Land	Daniel Meara		Daniel Meara
13	Land	Daniel Meara		Daniel Meara
14	Land	Mrs Mgt. Flannery		Mrs Mgt. Flannery
15	Land	John G. Mullany		John G. Mullany
16	Land	Patk Kelly, Thos. Keane, Thos. Cooney, Danl. Meara, Fergus Molloy		Patk Kelly, Thos. Keane, Thos. Cooney, Danl. Meara, Fergus Molloy
17	Land	Thos. Madden		Thos. Madden
18	Land	Jas. Shaughnessy		Jas. Shaughnessy

Townland: Cappasallagh, Co. Galway   Barony: Longford

No. on Plan	Description of Property	Owners	Lessees	Name and Address
1	Land	Richard C. Williams		Richard C. Williams
2	Land	Richard C. Williams		Richard C. Williams
3	Land	Richard C. Williams		Richard C. Williams

Townland: Cappasallagh, Co. Galway contd. Barony: Longford

No. on Plan	Description of Property	Owners	Lessees	Name and Address
4	Land	Richard C. Williams		Richard C. Williams
5	Land	John Coghlan		John Coghlan
6	Land	Michael Treacy (Boola)		Michael Treacy (Boola)
7	Land	Martin Fitzgerald		Martin Fitzgerald
8	Land	Francis Thomas Royston		Francis Thomas Royston
9	Land	Teresa Keane (wid.)		Teresa Keane (wid.)
10	Land	Thos. Meara		Thos. Meara
11	Land	Thos. Harris		Thos. Harris
12	Land	Nicholas Treacy		Nicholas Treacy
13	Land	Thos. Meara		Thos. Meara
14	Land	Mary Horan (wid.)		Mary Horan (wid.)
15	Land	Michael Treacy (Boola)		Michael Treacy (Boola)
16	Land	Thos. Harris		Thos. Harris
17	R.O.W.	Thos. Harris, Vincent Treacy, Thos. Scott, Nicholas Treacy, Thos. Meara, Teresa Keane (wid.)		
18	Land	Michael Treacy (Boola)		Michael Treacy (Boola)
19	Land	Jas. Callanan		Jas. Callanan
20	Land	Michael Treacy (Boola)		Michael Treacy (Boola)
21	Land	John Coghlan		John Coghlan
22	Land	Michael Fahy		Michael Fahy
23	Land	John Fahy		John Fahy
24	Land	Vincent Treacy		Vincent Treacy
25	Land	Thos. Scott		Thos. Scott
26	Land	Vincent Treacy		Vincent Treacy
27	Land	Michael Treacy (Boola)		Michael Treacy (Boola)
28	R.O.W.	Michael Treacy (Boola), Thos Harris, Delia Walsh (wid.), John Treacy, Wm Broderick, Michael Treacy, Nicholas Treacy, John Madden, Patk Dimond		
29	Land	Thos. Harris		Thos. Harris
30	Land	Delia Walsh (wid.)		Delia Walsh (wid.)
31	Land	John Treacy		John Treacy
32	Land	John Treacy		John Treacy

Townland: Cappasallagh, Co. Galway contd.   **Barony: Longford**

No. on Plan	Description of Property	Owners	Lessees	Name and Address
33	Land	John Treacy		John Treacy
34	Land	Wm Broderick		Wm Broderick
35	Land	Nicholas Treacy		Nicholas Treacy
36	Land	Michael Treacy		Michael Treacy
37	Land	Michael Treacy		Michael Treacy
38	Land	Nicholas Treacy		Nicholas Treacy
39	Land	John Madden		John Madden
40	Land	Patk Dimond		Patk Dimond

Townland: Derryhiveny South, Co. Galway   **Barony: Longford**

No. on Plan	Description of Property	Owners	Lessees	Name and Address
1	Land	Michael Donohue		Michael Donohue
2	Land	Rev. Fr. T. Porter PP		Michael Donohue
3	Land	Rev. Fr. T. Porter PP		William Kelly
4	Land	Patrick Meara		Patrick Meara
5	Land	Rev. Fr. T. Porter PP		William Kelly
6	Land	Patrick Meara		Patrick Meara
7	Land	Rev. Fr. T. Porter PP		William Kelly
8	Land	Patrick Meara		Patrick Meara
9	Land	Rev. Fr. T. Porter PP		William Kelly
10	Land	Patrick Meara		Patrick Meara
11	Land	John Smith		John Smith
12	Land	Rev. Fr. T. Porter PP		Michael Donohue
13	Land	Michael Donohue		Michael Donohue
14	Land	Rev. Fr. T. Porter PP		Thos. Harris
15	Land	Rev. Fr. T. Porter PP		Michael Treacy (Boola)
16	Land	Rev. Fr. T. Porter PP		John Kelly
17	Land	Rev. Fr. T. Porter PP		Jas. Cully
18	Land	Rev. Fr. T. Porter PP		John Smith
19	Land	Rev. Fr. T. Porter PP		Michael Daniels
20	Land	Rev. Fr. T. Porter PP		John Smith
21	Land	Rev. Fr. T. Porter PP		Daniel Caniff
22	Land	Rev. Fr. T. Porter PP		Michael Daniels
23	Land	Rev. Fr. T. Porter PP		Michael Daniels
24, 25	Land	Rev. Fr. T. Porter PP		John Smith, Daniel Caniff

Townland: Derryhiveny North, Co. Galway　　　　Barony: Longford

No. on Plan	Description of Property	Owners	Lessees	Name and Address
1	Land	Reps F.A.F. Seymour (Mrs Florence Seymour)		Reps F.A.F. Seymour (Mrs Florence Seymour)
2	Land	Reps F.A.F. Seymour (Mrs Florence Seymour)		Reps F.A.F. Seymour (Mrs Florence Seymour)
3	Land	Reps F.A.F. Seymour (Mrs Florence Seymour)		Reps F.A.F. Seymour (Mrs Florence Seymour)
4	Land	Reps F.A.F. Seymour (Mrs Florence Seymour)		Reps F.A.F. Seymour (Mrs Florence Seymour)
5	Land	Reps F.A.F. Seymour (Mrs Florence Seymour)		Reps F.A.F. Seymour (Mrs Florence Seymour)

Townland: Cappagh, Co. Galway　　　　Barony: Longford

No. on Plan	Description of Property	Owners	Lessees	Name and Address
1	Land	Thos. Fahy		Thos. Fahy
2	Land	Reps. Kate Flanagan (wid.)		Reps. Kate Flanagan (wid.)
3	Land	John Campbell		John Campbell
4	Land	Thos. Bohan		Thos. Bohan
5	Land	Michael McDonagh		Michael McDonagh
6	Land	Robt Tierney		Robt Tierney
7	Land	Chas Hogan		Chas Hogan
8	Land	Michael Ryan		Michael Ryan
9	Land	William Broder		William Broder
10	Land	John Gibbons		John Gibbons
11	Land	Martin Hayes		Martin Hayes
12	Land	Patk Nevin		Patk Nevin

Supplement No. 1 to Book of Reference No. 10

Townland: Keeloge, Co. Galway　　　　Barony: Longford

No. on Plan	Description of Property	Owners	Lessees	Name and Address
1	River Bank	Board of Works (BOW)		Commonage
9	Public Road	Galway Co. Council		Galway Co. Council
12	Yard	John Treacy		John Treacy
19, 20	Land	Charles Killeen		Charles Killeen

Supplement No. 2 to Book of Reference No. 10

Townland: Garrynatineel, Co. Tipperary   Barony: Owney & Arra

No. on Plan	Description of Property	Owners	Lessees	Name and Address
22, 23	Land	Samuel Whipp		D.L. Cooke
24 to 35 incl.	Land	Samuel Whipp		Samuel Whipp

■ Book of Reference No. 11

Townland: Tiranascragh, Co. Galway   Barony: Longford

No. on Plan	Description of Property	Owners	Lessees	Name and Address
1	Land	Margaret Keegan		Margaret Keegan
2	Land	Catherine Melody		Catherine Melody
3	Land	Patk Meheran		Patk Meheran
4	Land	Ml. Killacky		Ml. Killacky
5	Land	Martin Killacky		Martin Killacky
6	Land	John J. Moran		John J. Moran
7	Land	Patk Dermody		Patk Dermody
8	Land	John Slattery		John Slattery
9	Land	Peter O'Hara		Peter O'Hara
10	Land	Thos. Cooney		Thos. Cooney
11	Land	Reps. John Bigney (dec'd), (Martin Duane) (John Bohan Snr)		Reps. John Bigney (dec'd), (Martin Duane) (John Bohan Snr)
12	Land	John Bohan Snr		John Bohan Snr
13	Land	Thos. Keegan		Thos. Keegan
14	Land	Jos. Madden		Anne Madden
15	Land	William Hogan		William Hogan
16	Land	Anne Madden (wid.)		Anne Madden (wid.)
17	Land	Anne Madden (wid.)		Anne Madden (wid.)
18	Land	Lawrence Keegan		Lawrence Keegan

Townland: Meeneen, Co. Galway **Barony: Longford**

No. on Plan	Description of Property	Owners	Lessees	Name and Address
1	Land	John Taylor, Portumna		John Taylor, Portumna
2	Land	John Taylor, Portumna		John Taylor, Portumna
3	Lands & Marshy Drain	John Taylor, Portumna		John Taylor, Portumna
4	Land	John Taylor, Portumna		John Taylor, Portumna
5	Land & Drain	John Taylor, Portumna		John Taylor, Portumna
6	Land	John Horan Snr		John Horan Snr
7	Land	Thos. Mulhaire		Thos. Mulhaire
8	Rights of Way (Rs.O.W.)	John Horan Snr, Thos. Mulhaire, Patrick Burns, Michael Burns		
9	Land	John Horan Snr		John Horan Snr
10	Land	Michael Burns		Michael Burns
11	Land	Patrick Burns		Patrick Burns
12	Land	John Horan Snr		John Horan Snr
13	Land	Thos. Mulhaire		Thos. Mulhaire
14	Land	John Horan Snr		John Horan Snr

Townland: Muckanagh, Co. Galway **Barony: Longford**

No. on Plan	Description of Property	Owners	Lessees	Name and Address
1	Land	Thos. Grady		Thos. Grady
2	Land	Michael Burns		Michael Burns
3	Land	Michael Burns		Michael Burns
4	Land	Michael Burns		Michael Burns
5	Land	Michael Burns		Michael Burns
6	Land	William Forde		William Forde
7	Land	William Forde		William Forde
8	Land	William Forde		William Forde
9	Land	William Forde		William Forde
10	Land	William Forde		William Forde
11	Land	William Forde		William Forde
12	Land & bridge	Michael Burns		Michael Burns
13	Land	Michael Burns		Michael Burns
14	Land	John Horan Snr		John Horan Snr
15	Land	Michael Burns		Michael Burns
16	Land	Patrick Burns		Patrick Burns

Townland: Slaghta, Co. Galway   Barony: Longford

No. on Plan	Description of Property	Owners	Lessees	Name and Address
1	Land	William Forde		William Forde
2	Land	Jas. Dermody		Jas. Dermody
3	Land	John Dermody		John Dermody
4	Land	Thos. Cormican		Thos. Cormican
5	Land & Scrubby	Patrick Larkin		Patrick Larkin
6	Land	John Horan Jnr		John Horan Jnr
7	Land	John Horan Jnr		John Horan Jnr

Townland: Friars Island, Co. Galway   Barony: Longford

No. on Plan	Description of Property	Owners	Lessees	Name and Address
1	Land	Thos. Glynn		Thos. Glynn
2	Land	John Burns		John Burns
3	Land	Thos. Horan Snr		Thos. Horan Snr
4	Land	Mary Colahan (wid.)		Mary Colahan (wid.)
5	Land	Anne Dermody (wid.)		Anne Dermody (wid.)
6	Land	John Dempsey		John Dempsey
7	Land	Reps. Wm Gallagher (dec'd)		Reps. Wm Gallagher (dec'd)
8	Land	Patk Caragher		Patk Caragher
9	Land	Thos. Horan Snr		Thos. Horan Snr
10	Land	Ml. Bermingham		Ml. Bermingham
11	Land	Anne Dermody (wid.)		Anne Dermody (wid.)
12	Land	John Dempsey		John Dempsey
13	Land	Reps. Wm Gallagher (dec'd)		Reps. Wm Gallagher (dec'd)
14	Land	Mary Colahan (wid.)		Mary Colahan (wid.)
15	Land	Thos. Horan Snr		Thos. Horan Snr
16	Land	John Burns		John Burns
17	Land	Thos. Glynn		Thos. Glynn
18	Land	Richard Howard		Richard Howard
19	Land	Richard Howard		Richard Howard
20	Land	Board of Works (BOW)		

Townland: Coorinch, Co. Galway   Barony: Longford

No. on Plan	Description of Property	Owners	Lessees	Name and Address
1	Land	Johanna Killeen (wid.)		Johanna Killeen (wid.)

Townland: Meelick, Co. Galway — Barony: Longford

No. on Plan	Description of Property	Owners	Lessees	Name and Address
1	Land	Johanna Killeen (wid.)		Johanna Killeen (wid.)

Townland: Friarsland, Co. Galway — Barony: Longford

No. on Plan	Description of Property	Owners	Lessees	Name and Address
1	Land	Most Rev. Dr Dignam DD		Rev. Fr. Martin J Leahy PP
2	Land	Most Rev. Dr Dignam DD		Rev. Fr. Martin J Leahy PP
3	Land	Most Rev. Dr Dignam DD		Rev. Fr. Martin J Leahy PP

Townland: Keeloge, Co. Galway — Barony: Longford

No. on Plan	Description of Property	Owners	Lessees	Name and Address
1	Land	John Treacy		John Treacy
2	Land	Johanna Killeen (wid.)		Johanna Killeen (wid.)
3	Land	John Killeen		John Killeen
4	Land	John Killeen		John Killeen
5	Land	John Quinn		John Quinn
6	Land	John Quinn		John Quinn
7	Land	BOW		John Quinn
8	Small house	BOW		BOW

Townland: Co. Galway — Barony: Longford

No. on Plan	Description of Property	Owners	Lessees	Name and Address
1	Island	Most Rev. Dr Dignam DD		ILLAUNABOY ISLAND Rev. Fr. Martin J Leahy PP
1	Island	Most Rev. Dr Dignam DD		ILLAUNAGOUGHAL ISLAND Rev. Fr. Martin J Leahy PP
1	Island	Most Rev. Dr Dignam DD		CROMWELL'S ISLAND Rev. Fr. Martin J Leahy PP
1	Island	BOW	Rev. Fr. Martin J. Leahy PP	BRINOGE ISLAND Rev. Fr. Martin J Leahy PP
1	Island	BOW	Rev. Fr. Martin J. Leahy PP	KILLEEN'S ISLAND Rev. Fr. Martin J Leahy PP
1	Island	BOW	Rev. Fr. Martin J. Leahy PP	YELLOW ISLAND Rev. Fr. Martin J Leahy PP
1	Island	BOW	Mrs Christina Kelly	ILLAUNFORD ISLAND John Treacy

Townland: Co. Galway contd.   **Barony: Longford**

No. on Plan	Description of Property	Owners	Lessees	Name and Address
1	Island	BOW	John Quinn	LONG ISLAND John Quinn
1	Island	BOW	Mrs Christina Kelly	ILLAUNKYLE ISLAND John Treacy

Townland: Islands in Shannon (1-19 incl.), Co. Galway   **Barony: Longford**

No. on Plan	Description of Property	Owners	Lessees	Name and Address
13	Island	BOW	Rev. Fr. Martin J Leahy PP	Rev. Fr. Martin J Leahy PP
14	Island	BOW	Rev. Fr. Martin J Leahy PP	Rev. Fr. Martin J Leahy PP
15	Island	BOW	Rev. Fr. Martin J Leahy PP	Rev. Fr. Martin J Leahy PP
16	Island	BOW		Francis Cullen
17	Island	BOW	John Quinn	John Quinn
18	Island	BOW	Mrs Christina Kelly	John Treacy
19	Island	BOW	John Quinn	John Quinn

Townland: Fairyhill, Co. Galway   **Barony: Longford**

No. on Plan	Description of Property	Owners	Lessees	Name and Address
29	Land	Mark F. Tuohy		Mark F. Tuohy
30	Land	Mary A. Madden		Mary A. Madden
31	Public Road	Galway Co. Council		Galway Co. Council

■ Book of Reference No. 12

Townland: Ballykeelaun, Co. Clare   **Barony: Tulla Lower**

No. on Plan	Description of Property	Owners	Lessees	Name and Address
37,38	Land	Reps. of Michael Curry (dec'd) (Bridget Curry (wid.)		Bridget Curry (wid.)
39	House & Out-offices	Reps. of Michael Curry (dec'd) (Bridget Curry (wid.)		Bridget Curry (wid.)

Townland: Ruanard, Co. Clare   Barony: Tulla Lower

No. on Plan	Description of Property	Owners	Lessees	Name and Address
19	Land	Col. H.M. Butler Miss F.M. Walsh		Col. H.M. Butler Miss F.M. Walsh

Townland: Kildoorus, Co. Clare   Barony: Tulla Lower

No. on Plan	Description of Property	Owners	Lessees	Name and Address
54,55,56	Land	John Gorman		John Gorman

Townland: O'Briensbridge, Co. Clare   Barony: Tulla Lower

No. on Plan	Description of Property	Owners	Lessees	Name and Address
103	Co. Road	Clare Co. Council		Clare Co. Council
104		Lt.Col. Bryan, Mrs Marie Vaughan, Mrs Louisa Plunkett Little-Hamilton, Dr Daniel H. Foley		Rev. R.P. Baylee
105,106	Land	Lt.Col. Bryan, Mrs Marie Vaughan, Mrs Louisa Plunkett Little-Hamilton, Dr Daniel H. Foley		Daniel Fitzgerald
107	Land	Lt.Col. Bryan, Mrs Marie Vaughan, Mrs Louisa Plunkett Little-Hamilton, Dr Daniel H. Foley		Lt.Col. Bryan, Mrs Marie Vaughan, Mrs Louisa Plunkett Little-Hamilton, Dr Daniel H. Foley

■ Book of Reference No. 13

Townland: Mountcatherine, Co. Clare   Barony: Tulla Lower

No. on Plan	Description of Property	Owners	Lessees	Name and Address
15	Land	John J. De Courcey		John J. De Courcey

Townland: Kildoorus, Co. Clare   Barony: Tulla Lower

No. on Plan	Description of Property	Owners	Lessees	Name and Address
54	Land	Synan Campbell		Synan Campbell

Townland: Ardnataggle, Co. Clare   Barony: Tulla Lower

No. on Plan	Description of Property	Owners	Lessees	Name and Address
33,34	Land	Michael Cooney		Michael Cooney
35	Land	Patrick Crotty		Patrick Crotty
36,37	Land	Reps. Michael Hastings		Reps. Michael Hastings
38	Land	John Hastings		John Hastings

Townland: O'Briensbridge, Co. Clare   Barony: Tulla Lower

No. on Plan	Description of Property	Owners	Lessees	Name and Address
103 to 106 incl.	Land	Miss Delia McKeogh		Miss Delia McKeogh

Townland: Friars Island (Killaloe), Co. Tipperary   Barony: Owney & Arra

No. on Plan	Description of Property	Owners	Lessees	Name and Address
1	Land	Samuel Whipp		Samuel Whipp

Townland: Shantraud, Co. Clare   Barony: Tulla Lower

No. on Plan	Description of Property	Owners	Lessees	Name and Address
1	Land	Board of Works (BOW)		BOW
2	Land	BOW		Grand Canal Co.
3	Land	BOW		Denis Crowe
4	Land	BOW		Dudley Fletcher
5	Land	BOW		Denis Crowe

■ Book of Reference No. 14

Townland: Gortaha, Co. Galway   Barony: Longford

No. on Plan	Description of Property	Owners	Lessees	Name and Address
19, 20	Land	Thos. Shaughnessy		Thos. Shaughnessy
21	Land	Reps. Maria Maher (Fergus Molloy)		Reps. Maria Maher (Fergus Molloy)
22	Land	Walter J. Joyce		Walter J. Joyce

Townland: Killaloe Bridge, Co. Clare

No. on Plan	Description of Property	Owners	Lessees	Name and Address
1	Road bridge	Clare Co. Council		Clare Co. Council

Townland: Killaloe Bridge, Co. Tipperary   Barony: Tulla Lower

No. on Plan	Description of Property	Owners	Lessees	Name and Address
1	Road bridge	Tipperary Co. Council		Tipperary Co. Council

■ Book of Reference No. 15

Townland: Cloonfadda, Co. Clare   Barony: Tulla Lower

No. on Plan	Description of Property	Owners	Lessees	Name and Address
89	Land	Miss Rebecca O'Callaghan		Miss Rebecca O'Callaghan
90	Land	Miss Rebecca O'Callaghan		Miss Rebecca O'Callaghan
91	Land	Miss Rebecca O'Callaghan		Miss Rebecca O'Callaghan
92	Road	Miss Rebecca O'Callaghan		Miss Rebecca O'Callaghan
93	Land	Miss Rebecca O'Callaghan		Miss Rebecca O'Callaghan
94	Land	Miss Rebecca O'Callaghan		Miss Rebecca O'Callaghan
95	Land	Miss Rebecca O'Callaghan		Miss Rebecca O'Callaghan
96	Land	John Sheedy		John Sheedy
97	Land	John Sheedy		John Sheedy
98	Land	John Sheedy		John Sheedy

Townland: Killestry, Co. Clare   Barony: Tulla Lower

No. on Plan	Description of Property	Owners	Lessees	Name and Address
1	Land	Thos. Hogan		Thos. Hogan
2	Land	Thos. Hogan		Thos. Hogan
3	Land	Dr Wm Courtney		Dr Wm Courtney
4	Land	Dr Wm Courtney		Dr Wm Courtney
5	Land	Mary Ellen Maloney (wid.)		Mary Ellen Maloney (wid.)
6	Land	Thos. Hogan		Thos. Hogan
7	Land	Thos. Hogan		Thos. Hogan

Townland: Moys, Co. Clare   Barony: Tulla Lower

No. on Plan	Description of Property	Owners	Lessees	Name and Address
9	Land	Dr Wm Courtney		Dr Wm Courtney
10	Land	Dr Wm Courtney		Dr Wm Courtney
11	Land	Dr Wm Courtney		Dr Wm Courtney
12	Land	Dr Wm Courtney		Dr Wm Courtney
13	Land	Dr Wm Courtney		Dr Wm Courtney
14	Land	Dr Wm Courtney		Dr Wm Courtney
15	Canal & Embankments	Shannon Navigation Co.		Shannon Navigation Co.
16	Storehouse	Shannon Navigation Co.	Anthony Mackay	Anthony Mackay
17	Land	Rt Rev. Bishop of Killaloe		Rt Rev. Bishop of Killaloe
18	Land	Rt Rev. Bishop of Killaloe		Rt Rev. Bishop of Killaloe
19	Land	Rt Rev. Bishop of Killaloe		Rt Rev. Bishop of Killaloe
20	Land	Rt Rev. Bishop of Killaloe		Rt Rev. Bishop of Killaloe
21	Land	Rt Rev. Bishop of Killaloe		Rt Rev. Bishop of Killaloe
22	Land	Rt Rev. Bishop of Killaloe		Rt Rev. Bishop of Killaloe
23	Land	Rt Rev. Bishop of Killaloe		Rt Rev. Bishop of Killaloe
24	Land	Rt Rev. Bishop of Killaloe		Rt Rev. Bishop of Killaloe
25	Land	Rt Rev. Bishop of Killaloe		Rt Rev. Bishop of Killaloe
26	Land	Rt Rev. Bishop of Killaloe		Rt Rev. Bishop of Killaloe
27	Land	Major Lefroy		Major Lefroy
28	Land	Major Lefroy		Major Lefroy
29	Land	Rt Rev. Bishop of Killaloe		Rt Rev. Bishop of Killaloe
30	Land	Rt Rev. Bishop of Killaloe		Rt Rev. Bishop of Killaloe
31	Land	Rt Rev. Bishop of Killaloe		Rt Rev. Bishop of Killaloe
32	Land	Rt Rev. Bishop of Killaloe		Rt Rev. Bishop of Killaloe
33	Land	Rt Rev. Bishop of Killaloe		Rt Rev. Bishop of Killaloe
34	House	Rt Rev. Bishop of Killaloe		Francis Deane
35	Land	Rt Rev. Bishop of Killaloe		Rt Rev. Bishop of Killaloe
36	Store	Rt Rev. Bishop of Killaloe		Rt Rev. Bishop of Killaloe
37	Land	Major Lefroy		Major Lefroy
38	Land	Major Lefroy		Major Lefroy
39	House	Major Lefroy		Daniel Graham
40	House & Out-offices	Shannon Navigation Co. Limerick		John Magee

Townland: Roolagh, Co. Tipperary **Barony: Owney & Arra**

No. on Plan	Description of Property	Owners	Lessees	Name and Address
1	Land	Patrick J. Ryan		Patrick J. Ryan
2	Land	Patrick J. Ryan		Patrick J. Ryan
3	Land	Patrick J. Ryan		Patrick J. Ryan
4	Land	Patrick J. Ryan		Patrick J. Ryan
5	Rly & embankment	G.S. Rly		G.S. Rly
6	Land	Denis Kenna		Denis Kenna
7	Land	Board of Works (BOW)		BOW

Townland: Ballina, Co. Tipperary **Barony: Owney & Arra**

No. on Plan	Description of Property	Owners	Lessees	Name and Address
1	Land	BOW		BOW
2	Rly & embankment	G.S. Rly		G.S. Rly
3	Public Road	Nenagh Rural District Co		Nenagh Rural District Co

Townland: Garrynatineel, Co. Tipperary **Barony: Owney & Arra**

No. on Plan	Description of Property	Owners	Lessees	Name and Address
36	Land	Samuel Whipp		Samuel Whipp
37	Land	Samuel Whipp		Samuel Whipp
38	Land	Samuel Whipp		Samuel Whipp
39	Land	Wm Boweman		Wm Boweman
40	Land	Wm Boweman		Wm Boweman
41	Rly & embankment	G.S. Rly		G.S. Rly
42	Land	Samuel Whipp		Samuel Whipp
43	Land	Samuel Whipp		Samuel Whipp
44	Dwelling, out-offices and yard	Samuel Whipp		Samuel Whipp
45	Land	Samuel Whipp		Samuel Whipp
46	Out-offices	Samuel Whipp		Samuel Whipp
47	Land	Samuel Whipp		Samuel Whipp
48	Land	Samuel Whipp		Samuel Whipp
49	Land	Samuel Whipp		Samuel Whipp
50	Land	Samuel Whipp		Samuel Whipp

Townland: Garrynatineel, Co. Tipperary contd.   Barony: Owney & Arra

No. on Plan	Description of Property	Owners	Lessees	Name and Address
51	Land	Samuel Whipp		Samuel Whipp
52	Land	Samuel Whipp		Samuel Whipp
53	Land	Samuel Whipp		Samuel Whipp
54	Land	Samuel Whipp		Samuel Whipp
55	Land	Samuel Whipp		Samuel Whipp
56	Land	Samuel Whipp		Samuel Whipp
57	House	Samuel Whipp	Edward Healy	Edward Healy
58	Land	Samuel Whipp	Edward Healy	Edward Healy
59	Land	Samuel Whipp	Wm Burke	Wm Burke
60	Land	Samuel Whipp	Wm Burke	Wm Burke
61	Land	Samuel Whipp	Wm Burke	Wm Burke
62	Land	Samuel Whipp	Wm Burke	Wm Burke
63	Land	Nenagh R.D.Co.	Wm Boweman	Wm Boweman
64	House	Nenagh R.D.Co.	Wm Boweman	Wm Boweman

Townland: Island in Shannon, Co. Tipperary   Barony: Owney & Arra

No. on Plan	Description of Property	Owners	Lessees	Name and Address
1	Island	Samuel Whipp		Samuel Whipp

■ Book of Reference No. 16

Townland: Portumna Demesne, Co. Galway   Barony: Longford

No. on Plan	Description of Property	Owners	Lessees	Name and Address
2 Land	Viscount Lascelles			Viscount Lascelles

Townland: Fairyhill, Co. Galway   Barony: Longford

No. on Plan	Description of Property	Owners	Lessees	Name and Address
29	Land	Mark F. Tuohy		Mark F. Tuohy
30	Land	Mary A. Madden		Mary A. Madden
31	Land	Mary A. Madden		Mary A. Madden
32	Land	Patrick Shiel		Patrick Shiel
33	Public Road	Galway Co. Council		Galway Co. Council

■ Book of Reference No. 17

Townland: Ardclooney, Co. Clare   Barony: Tulla Lower

No. on Plan	Description of Property	Owners	Lessees	Name and Address
88	Land, Tow path & ½ bed of river	Comm Public Works (CPW)		CPW
89	R.O.W.	Public		Public

Townland: Birdhill, Co.Tipperary   Barony: Owney & Arra

No. on Plan	Description of Property	Owners	Lessees	Name and Address
50	ROW	Public		Public
51	½ bed of river	Jeremiah O'Sullivan		Jeremiah O'Sullivan

Townland: Redwood, Co.Tipperary   Barony: Ormond Lower

No. on Plan	Description of Property	Owners	Lessees	Name and Address
1 to 14 incl.	Land	Irish Land Commission (ILC)		ILC

■ Book of Reference No. 18

Townland: Ballina, Co. Tipperary   Barony: Owney & Arra

No. on Plan	Description of Property	Owners	Lessees	Name and Address
1	Railway Embankment & Land	G.S.Rly		G.S.Rly

Townland: Cullenagh, Co.Tipperary   Barony: Owney & Arra

No. on Plan	Description of Property	Owners	Lessees	Name and Address
1	Land	Board of Works (BOW)		James McKeogh
2	Railway Embankment & Land	G.S.Rly		G.S.Rly
3	Railway Station out-offices & platforms	G.S.Rly		G.S.Rly
4	Land	G.S.Rly		G.S.Rly

Townland: Cullenagh, Co.Tipperary contd. **Barony: Owney & Arra**

No. on Plan	Description of Property	Owners	Lessees	Name and Address
5	Goods Shed	G.S.Rly		G.S.Rly
6	Goods Shed	G.S.Rly		G.S.Rly

Townland: Portland Little, Co.Tipperary **Barony: Lower Ormond**

No. on Plan	Description of Property	Owners	Lessees	Name and Address
1 to 10 incl.	Land	Irish Land Commission (ILC)		ILC
11	Plantation	ILC		ILC
12,13,14	Land	ILC		ILC
15	Plantation	ILC		ILC
16,17	Land	ILC		ILC
18	Plantation	ILC		ILC

Townland: Portland Island, Co.Tipperary **Barony: Lower Ormond**

No. on Plan	Description of Property	Owners	Lessees	Name and Address
1, 2	Land	ILC		ILC

Townland: Shantraud, Co.Clare **Barony: Tulla Lower**

No. on Plan	Description of Property	Owners	Lessees	Name and Address
6	Corn mills, out-offices & yards	BOW		John Bannatyne & sons
7	Canal Bank	BOW		BOW

Townland: Ballina, Co. Tipperary **Barony: Owney & Arra**

No. on Plan	Description of Property	Owners	Lessees	Name and Address
1	Railway embankment & land	G.S.Rly		G.S.Rly

Townland: Cullenagh, Co. Tipperary **Barony: Owney & Arra**

No. on Plan	Description of Property	Owners	Lessees	Name and Address
1	Land	BOW		James McKeogh
2	Railway embankment & land	G.S.Rly		G.S.Rly

Townland: Cullenagh, Co. Tipperary contd.   Barony: Owney & Arra

No. on Plan	Description of Property	Owners	Lessees	Name and Address
3	Rly Station, out-offices & platforms	G.S.Rly		G.S.Rly
4	Land	G.S.Rly		G.S.Rly
5	Goods shed	G.S.Rly		G.S.Rly
6	Goods shed	G.S.Rly		G.S.Rly

Townland: Portland Little, Co.Tipperary   Barony: Lower Ormond

No. on Plan	Description of Property	Owners	Lessees	Name and Address
1 to 10 incl.	Land	ILC		ILC
11	Plantation	ILC		ILC
12	Land	ILC		ILC
13	Land	ILC		ILC
14	Land	ILC		ILC
15	Plantation	ILC		ILC
16	Land	ILC		ILC
17	Land	ILC		ILC
18	Plantation	ILC		ILC

Townland: Portland Island, Co.Tipperary   Barony: Lower Ormond

No. on Plan	Description of Property	Owners	Lessees	Name and Address
1,2	Land	ILC		ILC

Townland: Sallow Island, Co.Tipperary   Barony: Lower Ormond

No. on Plan	Description of Property	Owners	Lessees	Name and Address
1	Island	ILC		ILC

Townland: Shantraud, Co. Clare   Barony: Tulla Lower

No. on Plan	Description of Property	Owners	Lessees	Name and Address
6	Corn mills, out-offices & yards	BOW		John Bannatyne & sons

Townland: Shantraud, Co. Clare contd.   Barony: Tulla Lower

No. on Plan	Description of Property	Owners	Lessees	Name and Address
7	Canal Bank	BOW		BOW

■ Book of Reference No. 19

Townland: Springfield, Co. Clare   Barony: Tulla Lower

No. on Plan	Description of Property	Owners	Lessees	Name and Address
1	Land	Thos. McCormack		Thos. McCormack
2	Land	Thos. McCormack		Thos. McCormack
3	Land	Thos. McCormack		Thos. McCormack
4	Land	Thos. McCormack		Thos. McCormack
5	Land	Thos. McCormack		Thos. McCormack
6	Land	John Browne		John Browne
7	Land	John Browne		John Browne
8	Land	John Browne		John Browne
9	Canal Bank	Board of Works (BOW)		Jas. Good

Townland: Cloonlara, Co. Clare   Barony: Tulla Lower

No. on Plan	Description of Property	Owners	Lessees	Name and Address
38	Land	Pat. Stritch		Pat. Stritch
39	Land	Pat. Stritch		Pat. Stritch
40	Land	Pat. Stritch		Pat. Stritch
41	Land	Pat. Stritch		Pat. Stritch
42	Land	John Hamilton		John Hamilton
43	Land	John Hamilton		John Hamilton
44	Land	Rep. Church Body		Rev. Dean Gillespie
45	Land	Rep. Church Body		Rev. Dean Gillespie
46	Canal Bank	BOW		Michael Hackett

Townland: Rimeroe, Co. Clare   Barony: Tulla Lower

No. on Plan	Description of Property	Owners	Lessees	Name and Address
1	Land	Norah Heffernan		Norah Heffernan
2	Land	Norah Heffernan		Norah Heffernan
3	Land	Norah Heffernan		Norah Heffernan

Townland: Rimeroe, Co. Clare contd.　　Barony: Tulla Lower

No. on Plan	Description of Property	Owners	Lessees	Name and Address
4	Land	Norah Heffernan		Norah Heffernan
5	Land	Norah Heffernan		Norah Heffernan

Townland: Cottage, Co. Clare　　Barony: Tulla Lower

No. on Plan	Description of Property	Owners	Lessees	Name and Address
1	Land	John Cordue		John Cordue
2	Land	John Cordue		John Cordue
3	Land	John Cordue		John Cordue
4	Land	Mich. McKeogh		Mich. McKeogh
5	Land	Daniel Byron		Daniel Byron

Townland: Doonass Demesne, Co. Clare　　Barony: Tulla Lower

No. on Plan	Description of Property	Owners	Lessees	Name and Address
1	Land	Wm Torpey		Wm Torpey
2	Land	Wm Torpey		Wm Torpey
3	Land	Mrs Sgt. McNamara		Mrs Sgt. McNamara
4	Land	Laurence Meany		Laurence Meany

Townland: Newgarden, Co. Limerick　　Barony: Clanwilliam

No. on Plan	Description of Property	Owners	Lessees	Name and Address
1	Land	J.F. Morley		J.F. Morley
2	Land	Limerick Corporation		Limerick Corporation

Townland: Prospect, Co. Limerick　　Barony: Clanwilliam

No. on Plan	Description of Property	Owners	Lessees	Name and Address
1	Land	Limerick Corporation		Limerick Corporation
2	Land	Limerick Corporation		Limerick Corporation

River Shannon Fisheries, Co. Limerick

No. on Plan	Description of Property	Owners	Lessees	Name and Address
	Prospect Fishery	Col. Powell & Miss Johnson		Col. Powell & Miss Johnson
	Newgarden Fishery	J.F.E. Morley		J.F. Morley

River Shannon Fisheries, Co. Clare

No. on Plan	Description of Property	Owners	Lessees	Name and Address
	Landscape Fishery	Hugh O'Brien Moran		Hugh O'Brien Moran
	Errina Fishery	Irish Land Commission (ILC)		ILC

Townland: Grove Island, City of Limerick   Barony: Limerick

No. on Plan	Description of Property	Owners	Lessees	Name and Address
1	Island	Hon. Col. Knox		Patrick O'Connor

Appendix 2

Accidents and Deaths on the Shannon Scheme 1925–1929

(collated from relevant files of Shannon Power Development and from newspaper reports)

1925

28 October	Patrick O'Sullivan, Clonlara, labourer, married.
	Horse hit gatepost - cart overturned – killed – no inquest.

1926

9 January	Dr Klemer, engineer, Germany.
	Fell off Railway Bridge, Longpavement. Rescued from river – dislocated shoulder.
3 March	19 crew of SS *Arabia*.
	Lost at sea in storm.
9 March	Patrick O'Reilly, Clonlara.
	Hit on side by steam dredger.
29 March	Daniel Moloney, Kiladysart.
	Daniel O'Dwyer, Caherdaniel, Kerry.
	William Byrne, Arklow.
	Train ran into them at Ardnacrusha. Byrne killed.
22 April	Harry Dynan, Ennis.
	Mark McDermott, Leitrim.
	Caught between buffers at Parteen. Dynan broke a leg. McDermott received spinal injuries and died later.
4 May	Bridget Walshe, Foynes. Typist with Siemens.
	Cyclist. Bus accident at Rosmadda. Died of injuries.
19 June	Labourer Cussane, Killaloe.
	Machinery fell on him at Parteen.
3 July	James Dunne, Castlebar.
	Drowned bathing at Parteen.
8 July	Thomas O'Shaughnessy, Thomondgate.
	Knocked down by train at Blackwater – broken leg and back injury – critical.
9 July	Leonhard Gernsbeck, Germany, fitter.
	Electrocuted at Clonlara – high tension wire.

12 August	Edward Molloy, Limerick.
	Trailer over body – broken leg.
17 August	Francis Hogan, assistant driver, Limerick.
	Fell under trailer at docks – killed instantly.
	Anton Muendeerfer (reported as Menerdoff), mechanic,
	Germany. Foot injury.
18 August	Michael Mulcahy
	Fell off lorry at Crescent – killed.
21 August	John Crimmins, Limerick.
	Broken ankle at Clonlara.
	Franz Schiel, engineer, Germany.
	Fell from crane at Ardnacrusha – severe head, back and arm injuries.
	James King, Limerick.
	James Fitzgibbon, Limerick.
	Leg injuries at Blackwater.
28 August	James Colbert
	Broken leg – O'Brien's Bridge – removed to Scariff Cottage Hospital.
6 September	Alfons Feuerstein, carpenter foreman, Germany.
	Died in Dublin – cancer.
	Patrick Connolly
	Struck by train at Clonlara – arm and leg injuries.
	William O'Halloran, Killaloe.
	Train over leg – amputated - died 6 weeks later.
7 September	John Cox, Limerick.
	Hit by truck – back injury.
15 September	Patrick Connolly, Clonlara.
	Struck by train at Clonlara – arm and leg injuries.
	Thomas Hayes, Ennis.
	Back injury – Blackwater.
16 September	James Shinners, Newport, train driver.
	Scalded to death at O'Brien's Bridge.
	Thomas Gallowey, Banagher, fireman on train.
	Badly scalded; same occasion. Survived.
29 September	Dominic Barro, Co. Limerick.
	Lost three toes – drilling machine.
	Christopher Counihan
	Broke leg – Clonlara.

10 November	John Murphy, Limerick, dock-worker.
	Unloading SS steamer – killed on the docks.
21 November	Thomas Whiteside, Monaghan, pump attendant
	Electrocuted – Parteen.
27 November	John Foran, Abbeyfeale.
	Caught by lorry trailer – arm and leg injuries.
8 December	Four workmen injured – train toppled over.
	Michael Barrett, Mayo, back and chest injuries - Three others minor injuries.
20 December	Patrick Keane, Lahinch.
	Train shutter fell on him at weir – died.
	John Corbett, arm broken in same accident.

1927

15 January	Martin Lenihan, Nenagh – broken leg.
	Timothy Trotter – internal injuries.
	Train derailed at O'Brien's Bridge.
26 January	James Everett
	Injured back – railway sleeper fell on him.
28 January	Christopher Heffernan, Patrickswell.
	Drowned at O'Brien's Bridge Weir.
2 February	Michael Cooper, Mayo.
	Died of natural causes.
17 February	Patrick Grogan, Galway.
	Broken leg at Ardnacrusha.
24 February	Augustine Ryan, Limerick.
	Fractured wrist.
	Frank Heffernan, Blackwater.
	Concussed.
8 March	Edward O'Doherty, Kilmallock.
	Hit by flying rock, concussed.
24 March	Michael Dooley, New Ross.
	Injured in blasting operation at Parteen.
26 March	James Cuneen
	Caught in drilling machine at Ardnacrusha. Very seriously injured.
31 March	Martin Brennan, Tipperary.
	Spinal injuries – train buffers – Ardnacrusha.

2 April	John O'Brien, Cork.
	Leg injuries – electrical equipment exploded at O'Brien's Bridge.
23 April	Daniel Hayes, Galbally, Tipperary.
	Caught by train at Annegrove – killed.
4 May	John J. Hartnett, Limerick.
	Crushed by wagons at Ardnacrusha – killed.
	Patrick Cambell - fractured leg.
	Patrick O'Brien - fractured ankle.
7 May	Michael Hore, Limerick – fractured leg.
14 May	Joseph Slavin
	Injured tibia at O'Brien's Bridge.
17 May	Edward McCormack, Limerick.
	Killed by truck.
29 May	Johannes Buchberger, fitter, Germany.
	Died – Barrington's Hospital – pneumonia and heart failure.
6 June	Mr Patterson, Limerick.
	Severe electric shock – O'Brien's Bridge – lost arm and two feet.
16 June	Messrs Doyle.
	Dixon.
	Conron.
	Injured at Clonlara by excavator.
18 June	Edward Barry
	Head injury – excavator – Clonlara.
20 June	James Martin, Tipperary.
	Caught in hawser – multiple injuries.
25 June	Thomas Kiely, Galway.
	Severely injured by railway bogey.
5 July	Edward Prendeville
	Rock fell on him. Survived.
8 July	Paul Berlin, excavation foreman, Germany.
	Died in Barrington's Hospital – Addison's Disease.
16 July	Fritz Zaum, mechanic, Germany.
	Injured by bogey at O'Brien's Bridge.
23 July	Mrs Flynn of Davidson's Bar.
	Shutter fell on her – killed at Parteen.

6 August	Joseph Toohill
	Knocked down at Larkin's Cross by Siemens' car.
20 August	Denis Sheridan
	Caught in concrete mixer and severely crushed at Parteen.
26 August	Labourer Doherty
	Railway irons fell on him.
27 August	John Howard, Thurles.
	Rails fell on him – multiple injuries.
1 September	Patrick Costelloe, Limerick.
	Christopher Collins, Dublin.
	Injured at Ardnacrusha – legs and back.
15 September	Thomas O'Brien
	Caught in concrete mixer – O'Brien's Bridge – compound fracture.
25 October	Tim Hayes
	Pole fell on him at Parteen – thigh injury.
	Patrick Whelan
	Drowned in cutting of SS, Clonlara.
11 November	Garda Jones, Clonlara
	Injured by flying timber during blasting operation.
26 November	James Enright, Shanagolden.
	Hand injury – Clonlara.
12 December	Patrick O'Shaughnessy
	Fractured leg, Annegrove.
13 December	John O'Reilly.
	Legs injured by bogey wagons at Ardnacrusha.
	James Courtney, Killaloe.
	Killed by rockslide while excavating at Parteen.
29 December	James Mulligan, Kerry.
	Fell into quarry at O'Brien's Bridge – spinal injury.

1928

12 January	Patrick Timmons
	Fractured leg – blasting operation.
	James Collins, Newtown.
	Drowned in canal at Killaloe – missing for some time.

20 January	Patrick Wyse, Limerick.
	Killed by trailer of lorry, Blackwater.
28 February	William McGrath, Roscrea.
	Run over by bus – Ardcloney Bridge.
4 March	James Heery, mechanic, Portumna.
	Died when compressor exploded.
10 March	John Gavin, dockworker.
	Fell into hold of ship – *SS Steiner*
31 March	Linus Krieger, machinist, Germany.
	Anton Grossman, electrical fitter, Germany.
	Crane toppled – trapped them. Grossman seriously injured.
28 April	Michael Keogh, Killaloe.
	Fell between trains – injured.
1 May	Jerry Keane, Limerick.
	John Keane, Limerick.
	Brothers, fell thirty feet into quarry at Ardnacrusha – head, chest and leg injuries.
3 May	Fritz Finck, engineer.
	Fell into rock fissure at O'Brien's Bridge – spinal injury and broken arm and leg
5 May	Margaret Taylor, Blackwater.
	Killed by truck trailer at The Huts, Blackwater.
17 May	Patrick McMullan, Prospect Depot.
	Girder fell on him.
22 May	William Buckley, Dublin.
	Legs injured by train buffers.
29 May	John Touhy
	Wagon tipping accident at Clonlara – broken leg.
2 June	Thomas O'Sullivan, Galway.
	Leg crushed by machine at Blackwater.
19 June	Carol Mullins
	Head injuries – unloading wagon – O'Brien's Bridge
26 June	James Barrett
	Caught in machinery at Ardnacrusha – arm, chest and back injuries.
24 July	Martin Joyce, Castleblaney, Monaghan.
	Drowned in Shannon at Parteen.

25 July	Labourer Breary
	High-voltage shock at Ardnacrusha – not injured!
9 August	Bartholomew Fahey
	Head injuries from train wagon at Parteen.
11 August	John Fahy
	Head injuries – train buffers.
	James Markham, Thomondgate.
	Electrocuted at Clonlara.
15 August	Herman Gunther, granite worker, Germany.
	Injured both feet, broken ankle – rubble fell on him.
	David Brennan
	Bruised thigh – same occasion.
16 August	Patrick Carolan, Drogheda.
	Killed tilting wagons at Weir.
21 August	Chris Lyons, Limerick
	Patrick Cronin, Galway.
	Both injured by flying pieces of iron at Clonlara.
30 August	John Dundon
	Fractured leg – falling railway sleepers at O'Brien's Bridge.
	John Dooley
	Back injury – falling railway sleepers at O'Brien's Bridge.
23 September	Frank Murphy, Tralee.
	Drowned from boat at O'Brien's Bridge.
4 October	Matthaeus Morlock, dredger driver, Germany.
	Fell from crane – fractured leg, head injuries at O'Brien's Bridge.
12 October	Josef Horner, excavator foreman, Germany.
	Killed at Parteen – explosives.
22 October	Rudolf Weigel, carpenter, Germany.
	Fell thirty-three feet at Intake Building, Ardnacrusha – killed instantly.
3 November	Frank King, Limerick.
	Fell into quarry, broken leg.
18 November	Denis Healy, Kilfinane.
	Patrick Folan, Galway.
	Daniel O'Brien, Limerick City.
	All three killed tilting wagon at Clonlara.

20 November	Patrick Brinn, Limerick.
	Wire rope broke on trestle at Parteen – killed in fall.
	Patrick Tracy, Inch.
	Broken leg – same incident.
27 November	Patrick Larkin, Dublin.
	Fell in turbine emplacement – back injuries.
28 November	Jim McMullen, Down.
	Injured in fall at Ardnacrusha.
11 December	John Ryan, Castleconnell.
	Killed by bus at O'Connors' Cross, going to work at Clonlara.
21 December	Jacob Kunz, Germany, foreman.
	Murdered on railway line - Parteen.
27 December	Daniel O'Brien, Tipperary.
	Run over by train at Parteen. Had served in Flying Corps in Great War.

1929

9 January	James Walshe, Nenagh.
	Head crushed by wagons at O'Brien's Bridge.
	Michael Ryan
	Ear injury – same occasion.
16 January	James King, Clifden, Galway.
	Drowned at O'Brien's Bridge.
26 January	Edward Nolan
	Fell thirty feet – head and thigh injuries.
16 February	John O'Keeffee
	Injured knee – train buffers – at O'Brien's Bridge.
20 February	Martin Doherty, Mayo.
	Fell 18 feet from stone crusher, injured head, legs and arms.
23 February	David Lyons
	Fell 35 feet into tailrace – badly injured.
20 March	John Gubbins
	Broken leg – Clonlara.
	John O'Reilly
	Broken leg – O'Brien's Bridge.

30 March	Edward Fahy
	Fell off tank at Ardnacrusha, broke both ankles.
17 April	John Sheehan
	Edward McElligott
	Train tipping accident at Clonlara.
27 April	John Sheedy
	Train tipping injury.
6 May	Dermot McDonagh, Galway.
	Knocked down by train at O'Brien's Bridge
16 May	John J. Larkin, Clonfert.
	Joseph Twoomey, Kilmallock.
	Both killed when part of dredger fell on them at weir.
7 June	John Rigney, Laois.
	Killed by tip wagon.
17 June	Denis Hishon, Charleville.
	Injured leg at Ardnataggle.
29 June	William Maloney, Broadford.
	Injured knee – tilting wagon.
16 July	Dominic Barron
	Injured arm – struck by bogey.
29 July	Tim Dennehy
	John Taylor
	Injured at Power Station.
12 August	Joe O'Leary
	Patrick Ahearne
	John O'Sullivan
	Injured by bogey wagons at O'Brien's Bridge.
14 August	Peter Kinsella, Kildare.
	Michael Conroy, Kildare.
	Fell off scaffolding at O'Brien's Bridge.
2 October	Ernest Jacob, Germany.
	Fell twenty feet off ladder at Ardnacrusha – injured legs.
6 October	Patrick Moore, Tipperary.
	Electrocuted at Power Station.

This list of accidents is by no means exhaustive, as many more went unreported in the newspapers and unrecorded in the official files of Shannon Power Development. This is understandable as all of the claims for compensation under the various Workmen's Compensation Acts were made against the contractors, Siemens Bauunion, Siemens Schuckert and the ESB, thereby excluding the involvement of Shannon Power Development. The registers of awards in the courts are of interest because many new names appear as well as the amounts awarded.

The relevant list of cases and awards heard in Ennis Court:

1.	15 May 1926	John Horkan, Castlebar, plumber - £46 5s.
2.	18 May 1926	Patk. Carey, Limerick - £25.
3.	"	Lizzie O'Brien, Arklow - £120.
4.	"	James Ryan, O'Brien's Bridge -£20.
5.	25 June 1926	Michael Pickford, Ennis - £50.
6.	"	Thomas Silke, Limerick - £70.
7.	9 July 1926	Francis Burke, Cappamore - £20.
8.	"	Joseph Ryan, Limerick - £35.
9.	3 September 1926	Martin O'Connell, Limerick - £20.
10.	7 September 1926	Edward McAuliffe, Bruff - £15.
11.	11 September 1926	George McHugh, O'Brien's Bridge - £20.
12.	16 September 1926	William Woulfe, Parteen - £15.
13.	"	James O'Connor, Kilkee - £15.
14.	17 September 1926	John and Mary McDermott, Leitrim - £50.
15.	18 September 1926	John Fitzpatrick, Killaloe - £15.
16.	29 September 1926	Martin Thynne, Miltown Malbay - £100.
17.	8 October 1926	George James O'Sullivan, Killaloe - £55.
18.	15 October 1926	Michael Warren, Kilmaley - £50.
19.	19 October 1926	Thomas O'Connell, Limerick - £20.
20.	5 November 1926	Edward Molloy, Limerick - £30.
21.	"	Richard Shinners et al. - £300.
22.	16 November 1926	Timothy Ryan, Lisnagry - £15.
23.	18 November 1926	Patrick Hayes, Killaloe - £30.
24.	3 December 1926	Denis Gossane, Killaloe - £100.
25.	7 December 1926	Michael Dynan, Ennis - £50.

26.	9 December 1926	James Gallagher, Limerick - £30.
27.	14 December 1926	John O'Shea, Hospital - £40.
28.	22 December 1926	Thomas Kileen, Ballyhaunis - £50.
29.	5 January 1927	Thomas Lyons, Tipperary Town - £30.
30.	8 January 1927	Thomas Sheehan, Ardfert - £30 10s.
31.	10 January 1927	James Hewitt, Newport – Application refused with costs.
32.	13 January 1927	John Keane, Lahinch - £75.
33.	"	James O'Brien, Limerick – Dismissed.
34.	21 January 1927	John Whiteside et al., Monaghan - £100.
35.	"	Patrick Connolly, Clifden - £15.
36.	27 January 1927	Daniel Mahony, O'Brien's Bridge - £60.
37.	"	James Fenelon (Kehoe), Wexford – No Appearance.
38.	11 February 1927	Alphonsus McGannon, Quilty - £40.
39.	"	John Corbett, Killaloe - £75.
40.	"	John Madden, Galway - £35.
41.	14 April 1927	Catherine Heffernan, Askeaton - £150.
42.	13 May 1927	Michael Dynan, Ennis - £30.
43.	"	Michael and Kate O'Halloran, Killaloe - £160.
44.	10 June 1927	Franz Schiel, Germany -£350.
45.	23 June 1927	John Shaughnessy, Killaloe - £100.
46.	6 July 1927	Thomas and Mary Anne Hayes, Drumcureen - £10.
47.	18 July 1927	Timothy Murphy, Cahermaine - £50,
48.	31 August 1927	William Ryan, Annacotty - £20.
49.	14 September 1927	James Kearney, Blackwater - £20.
50.	"	John Tracy, Limerick - £40.
51.	"	Michael Kiely, Limerick - £15.
52.	15 September 1927	William Chamberlain, Caherconlish - £20.
53.	27 September 1927	Patrick O'Shaughnessy, Thomondgate - £760.
54.	10 October 1927	Michael Browne, Dublin - £75.
55.	4 November 1927	Peter Boyle, Ardara, Donegal - £30.
56.	"	James Doherty, Drumkenelly, Donegal - £35.
57.	"	Laurence Loftus, Foxford, Mayo – £30.
58.	28 November 1927	Francis Brennan, Dungloe, Donegal - £40.
59.	20 December 1927	Thomas Keely, Rosmadda -£25.
60.	7 January 1928	James Henebry, Blackwater Rd. - Dismissed.

61.	23 January 1928	John McHale, Knockmore, Mayo - £40.
62.	26 January 1928	Timothy McCarthy, Ardnacrusha – Refused with costs.
63.	20 February 1928	Timothy Crotty, O'Brien's Bridge - £120.
64.	2 March 1928	Terence O'Brien, Dublin - £30.
65.	6 March 1928	John Hayes, Dublin - £25.
66.	10 March 1928	Lily Heery, Portumna - £300.
67.	13 March 1928	Colman Nee, Lettermore - £20.
68.	"	Patrick Geoghegan, Inverin - £50.
69.	"	Dudley Joyce, Costelloe, Galway - £100.
70.	"	Thomas Naughton, Costelloe - £100.
71.	"	Thomas Geoghegan, Inverin - £12.
72.	27 March 1928	James O'Gorman, Nenagh - £76 10s.
73.	29 March 1928	Michael Dooley, New Ross - £250.
74.	30 March 1928	Patrick Casey, Killaloe - £17 10s.
75.	5 April 1928	Patrick O'Brien, Thurles - £60.
76.	12 April 1928	Denis Madden, Nenagh - £50.
77.	"	John Fitzpatrick, Killaloe – Refused with costs.
78.	25 April 1928	Patrick Gillmar, Mountrath - £100.
79.	28 April 1928	Michael Nihill, Sixmilebridge - £40.
80.	10 May 1928	James Kavanagh, Ardnacrusha - £15.
81.	13 July 1928	Michael Conway, Killaloe - £25.
82.	31 July 1928	Patrick O'Flaherty, Cobh - £20.
83.	"	James Sexton, Cobh - £215.
84.	22 August 1928	Frank Bennett, Tipperary town - £25.
85.	4 September 1928	Michael Nee, Lettermore - £50.
86.	"	Thomas Conway, O'Brien's Bridge - £25.
87.	24 October 1928	Geoffrey O'Connell, Blarney - £30.
88.	19 January 1929	Timothy Ryan, Tipperary town - £26 10s.
89.	9 August 1929	James Kavanagh, Dublin – Discontinued/Nil.
90.	28 November 1929	Bernard Rodgers, Dublin - £22.
91.	6 October 1931	Patrick Aherne, O'Brien's Bridge - £650.
92.	18 March 1932	Thomas O'Brien, O'Brien's Bridge - £100.
93.	14 April 1932	Paul Prendergast, Blackwater – Withdrawn By Leave.
94.	25 November 1932	Timothy O'Donnell, Limerick - £50.

95.	16 December 1932	James Ryan, Blackwater - £20.
96.	2 March 1933	James Griffin, Gillogue £35.
97.	26 June 1933	Thomas Myers, Oatfield - £44 18s.
98.	8 September 1933	Patrick Kelleher, Corbally - £40.
99.	29 September 1933	Elisabeth Coonan and children, Killaloe - £300.

The National Archives number for the Clare Register is 1D 14 77.

The Limerick courts were equally busy adjudicating on applications for compensation under the Workmen's Compensation Act. The number of the relevant Registers are – 1D 42 143/4/5. The style of entry is different to the Clare Register however, in that the actual date of the accident is noted but the person's address is omitted.

The following is the list of entries from the Court Minute Books relating to accidents on the Shannon Scheme:

1.	10 June 1926	Patrick Markham - £1 5s. for six months.
2.	29 May 1926	John McElligott - £30.
3.	8 November 1926	Richard Roberts – Application refused.
4.	"	Catherine Hogan (widow) - £283.
5.	"	Daniel Mulcair - £30.
6.	"	Patrick Hogan - £75.
7.	8 February 1927	Thomas Mason – Withdrawn.
8.	"	Margaret Murphy (widow) - £300.
9.	25 May 1927	John William Cooke - £150.
10.	"	Mother of Michael Hartnett - £100.
11.	"	Babette Gernsbeck - £300.
12.	20 July 1927	Mary Anne McCormack – Appealed.
13.	8 November 1927	John Galligan – case settled (no sum entered).
14.	"	Patrick O'Reilly - £250.
15.	"	John Butler – £24.
16.	20 February 1928	Michael Daly - £27 14s. 2d.
17.	21 February 1928	Michael Davern - £20 15s. 4d.
18.	"	John O'Reilly – Order refused.
19.	"	William O'Grady – Order Refused.
20.	"	Patrick Donohoe - £31 10s.

21.	"	William Brunnock - £20.
22.	"	Martin Lenihan - 15s. per week until review.
23.	27 February 1928	Christine Courteney (widow) - £300.
24.	1 June 1928	Michael Daly – Review of weekly payment - £20.
25.	"	Patrick Powell – 15s.
26.	"	Denis Cagney - £23 13s. 4d.
27.	"	Myles Kenny - £10.
28.	"	Michael McGuire – Struck out.
29.	"	Owen O'Donnell – Adjourned.
30.	"	Michael Troy - £27.
31.	"	Thomas Hiskin - £25.
32.	13 July 1928	Owen O'Donnell - £10.
33.	"	Joseph Martin - £50.
34.	"	James McKelvey - £33 8s.
35.	9 November 1928	Reginald Gilligan - £1 15s. for three weeks.
36.	"	Patrick Campbell - £50.
37.	"	Denis Goode - £20.
38.	"	John Herron – Refused with costs.
39.	"	Thomas Conway – Not concluded.
40.	"	Thomas O'Halloran - £110.
41.	"	Michael Considine - £55.
42.	"	John Roche - £5.
43.	"	Stephen Markham – Refused.
44.	"	Margaret Murphy - £50.
45.	4 December 1928	Ellen Ryan (applicant for Daniel O'Brien) - £300.
46.	18 December 1928	Patrick O Donohue – Review £240 compensation.
47.	"	Martin Lenihan – 15s. reduced to 3s. 6d. per week.
48.	"	John McKelvey – Review. Award ceases 1 July 1929.
49.	"	Michael Ryan - £20.
50.	"	John J. Kilmartin - £50.
51.	"	Patrick Healy - £100.
52.	"	Margaret Brinn (widow) - £300.
53.	"	John Duggan – Settled – no details entered.
54.	"	Anton Grossman – Agreement signed.
55.	"	John Noble - £20.

56.	15 May 1929	James Lalor - £25.
57.	"	Christopher Godfrey - £40.
58.	"	Francis Byrne – Refused with costs.
59.	"	Edward Connors – 15s.
60.	"	Martin Walshe – Refused with costs.
61.	"	William Walshe - £50.
62.	"	Patrick Healy - £100.
63.	"	Ellen O'Brien - £100.
64.	"	William Nugent – Struck out.
65.	"	Michael Cooney – Settled – no details.
66.	"	Patrick J. Gallagher – 1d. per week for 12 weeks.
67.	"	Timothy Hayes - £28.
68.	19 July 1929	Martin Walshe – Postponed.
69.	"	Alexander Goodwin – Refused.
70.	"	Frank King - £15 15s.
71.	"	Thomas Fahy - £226 5s.
72.	"	William Byrne - £5 12s. 6d.
73.	"	William Ryan – 15s. per week.
74.	"	Mary Ellen Larkin - £200.
75.	"	John (Martin) Folan - £200.
76.	"	Ellen Twoomey (widow) - £200.
77.	11 November 1929	Martin Walshe – Refused.
78.	"	John Hafford – Refused.
79.	"	James Keane - £15.
80.	"	John Walshe - £15.
81.	"	John Sheehan – Refused with costs.
82.	"	Patrick Murphy - £20.
83.	"	John Grace – Refused with costs.
84.	12 November 1929	Thomas Weldon – £18.
85.	"	John Dundon – £24 10s.
86.	"	Daniel Lynch – Application struck out.
87.	"	James O'Neill – 2s. 6d. per week reduced to 6d. per week.
88.	"	Thomas Kinsella - £50.
89.	"	Denis Hiskin – Refused.
90.	"	Dominic Barron - £1 16s. for two weeks.

91.	"	Herman Gunther – Refused with costs.
92.	"	Christina Courteney - £172 2s. 1od.
93.	23 November 1929	William Ryan - £15 15s.
94.	17 November 1929	Patrick Dolan - £150.
95.	3 February 1930	Edward Connors – Award of 5s. per week to cease.
96.	"	William Byrnes – Review and Terminate – Withdrawn by applicant.
97.	"	John Dundon – Review and Terminate – Refuse application to review.
98.	"	John Hernon - Now over 21 - £50 15s. 8d.
99.	"	Sarah Moore (widow) - £297 13s. 4d.
100.	"	Thomas Kinsella -£50 4s. 4d.
101.	"	John Hernon – Adjourned.
102.	29 April 1930	John Hernon – Refused with costs for this hearing.
103.	"	Timothy Long – Review – Award reduced to 1d. per week.
104.	"	Michael Broderick - £8 6s. 8d.
105.	"	Martin Fitzgerald - £19 5s.
106.	"	Michael Brown - £6 2s. 6d.
107.	"	Patrick Murphy - £15.
108.	"	Ellen and Mary Rigney - £175.
109.	"	Timothy Long – Review and increase – 1os. per week.
110.	"	Timothy Hayes – Review and reduce – 18s. per week.
111.	"	Christopher O'Connell -£18 19s. 2d.
112.	"	William Ryan – Review and reduce – 15s. to 1os. per week.
113.	"	Ellen Twoomey - £200.
114.	"	Ellen Rigney - £15.
115.	12 November 1930	Stephen Colbert - £25.
116.	"	David Lonergan - £50.
117.	"	Ellen Rigney - £1 7s. 1od. interest.
118.	"	Ellen Twoomey - £50 2s.
119.	"	Christopher O'Connell – Refused.
120.	"	Denis J. Daly - £30 6s. 8d.

Appendix 3

List of German Workmen (NA, SS13928)

No:	Surname	Christian Name	Occupation	Date of Arrival	Date of Return	Existing Permit Expires
1	Apelt	Robert	Foreman	15.10.25	15.11.27	22.1.27
2	Backes	Paul	Carpenter	29.10.25	25.4.29	No time condition
3	Baldenhofer	Paul	Carpenter	29.10.25	31.10.29	No time condition
4	Bamberger	August	Electrical Fitter	29.10.25	4.10.29	22.11.26
5	Behrendt	Georg	Machinist	22.1.26	30.4.26	-------
6	Bergbauer	Johann	Dredger Foreman	21.9.25	23.6.29	12.1.27
7	Bergemann	Erich	Mechanic	22.1.26	31.3.26	-------
8	Berghoff	Paul	Electrical Fitter	10.12.25		10.12.26
9	Berheide	Albert	Machine Erector	22.1.26	20.6.26	-------
10	Bieber	Friedrich	Carpenter	29.10.25	27.9.26	No time condition
11	Bermann	Julius	Machinist	6.1.26	8.8.26	-------
12	Brueggemann	Heinrich	Smith	29.10.25		10.12.26
13	Brumme	Karl	Machine Erector	6.1.26	2.10.26	15.10.25
14	Buchberger	Hans	Electrical Fitter	22.1.26	@30.5.27	22.11.26
15	Buehler	Otto	Foreman	21.9.25	9.10.28	12.1.27
16	Burkhardt	Adolf	Mechanic	1.11.25	17.5.27	1.11.26
17	Burkhardt	Hermann	Foreman for drilling	13.12.25	15.12.27	22.5.27
18	Burkhardt	Max	Mechanic	21.9.25	17.2.28	12.1.27
19	Behrendt	Hermann	Mechanic	2.2.26	23.6.29	22.2.27
20	Bogenhuber	Fritz	Concrete Specialist	2.2.26	7.9.28	22.5.27
21	Braun	Konrad	Store Administrator	2.2.26	31.3.27	No time condition
22	Bretz	Wilhelm	Mechanic	2.2.26	15.8.29	22.11.26
23	Brueck	Ludwig	Foreman	2.2.26	11.12.28	22.5.27
24	Broecher	Fritz	Machinist	5.2.26		22.2.27
25	Boehlicke	Max	Machine Erector	12.2.26	26.6.26	-------
26	Britt	Franz	Machine Erector	23.2.26	9.7.26	-------
27	Buendgens	Johann	Foreman	1.3.26	26.6.29	22.5.27
28	Becher	Robert	Foreman	9.3.26	27.9.26	22.5.27
29	Buettner	August	Dredger Foreman	9.3.26	25.4.29	10.12.26
30	Babl	Franz	Foreman	26.3.26	27.7.29	26.3.27
31	Chiri	Ludwig	Electrical Fitter	9.3.26	11.3.28	10.12.26
32	Dahmes	Max	Machine Erector	15.1.26		-------

33	Dexel	Albert	Foreman	13.12.25	24.8.29	13.12.26
34	Diehm	Georg	Carpenter	1.11.25		1.11.26
35	Diehm	Kaspar	Foreman	21.9.25	4.10.29	12.1.27
36	Dierck	Hermann	Machine Erector	10.12.25	14.9.27	——-
37	Dierkopf	Max	Carpenter	15.1.26	27.9.26	22.5.17
38	Dumm	Joseph	Mechanic	1.11.25	20.2.26	——-
39	Dynovski	Willi	Joiner	1.11.25	27.9.26	No time condition

– 2 –

No:	Surname	Christian Name	Occupation	Date of Arrival	Date of Return	Existing Permit Expires
40	Draeger	Fritz	Machine Erector	2.2.26	12.12.26	15.10.26
41	——-					
42	Dettenbach	Franz	Foreman	26.3.26		26.3.27
43	Eschenback	Walter	Mechanic	5.10.25		12.1.27
44	Eslinger	Fritz	Machine Erector	13.12.25		1.1.27
45	Eickworth	Max	Engine Driver	2.2.26	15.4.26	——-
46	Edelmann	Gustav	Dredger Foreman	9.3.26	30.5.27	10.12.26
47	Englert	Joseph	Dredger Foreman	9.3.26	25.4.29	10.12.26
48	Fath	Joseph	Dredger Foreman	21.9.25	27.2.29	12.1.27
49	Feist	Otto	Mechanic	1.11.25	29.3.28	1.11.26
50	Feuerstein	Alfons	Carpenter Foreman	5.10.25	22.6.28	13.1.27
51	Feuerstein	Alfred	Mechanic	13.12.25	24.8.29	22.11.26
52	Fink	Fritz	Foreman	5.10.25	22.6.28	13.1.27
53	Finke	Friedrich	Carpenter	1.11.25	27.9.26	10.12.26
54	Finze	Otto	Dredger Foreman	10.12.25	16.3.28	101.12.26
55	Floerke	Fritz	Mechanic	1.11.25	20.2.26	——-
56	Frank	Joseph	Mechanic	21.9.25	28.11.26	12.1.27
57	Frege	Berharnd	Mechanic	22.1.26	25.10.28	22.11.26
58	Freihold	Hugo	Machine Erector	10.12.25		10.12.26
59	Frenzel	Johannes	Electrical Fitter	22.1.26		22.11.26
60	Frese	August	Mechanic	21.9.25		12.1.27
61	Fritzsche	Georg	Mechanic	4.10.25		No time condition
62	Frommhold	Moritz	Mechanic	1.11.25	19.2.29	1.11.26
63	Fuehr	Georg	Electrical Fitter	15.1.26	5.4.28	22.11.26
64	Feitzik	Heinrich	Mechanic	2.2.26		22.11.26
65	——-					
66	Fischer	Erich	Student	24.3.26	31.7.26	——-
67	Gause	Walter	Motor Driver	5.10.26	22.3.26	——-

68	Geisler	Otto	Carpenter	1.11.25	25.6.29	No time condition
69	Goertz	Adolf	Carpenter Foreman	15.1.26		22.5.27
70	Goldbeck	Otto	Carpenter	1.11.25	27.9.26	——-
71	Goldhuber	Franz	Carpenter	1.11.25	27.9.26	No time condition
72	Gollenstede	Georg	Miner	6.1.26	24.8.29	22.5.27
73	Graeber	Peter	Machinery Foreman	10.12.25	15.8.29	10.12.26
74	Greil	Hermann	Mechanic	21.9.25		12.1.27
75	Griese	Hans	Carpenter Foreman	1.11.25	27.9.26	No Time condition
76	Grossmann	Anton	Electrical Fitter	12.9.25	24.8.29	12.1.27
77	Guhl	Richard	Carpenter	1.11.25	27.9.26	No time condition
78	Guettel	Karl	Carpenter	1.11.25	27.9.26	No time condition
79	Gundlach	August	Carpenter	1.11.25	27.9.26	10.12.26

– 3 –

No:	Surname	Christian Name	Occupation	Date of Arrival	Date of Return	Existing Permit Expires
80	Gierth	Wilhelm	Engine Driver	5.2.26	5.11.28	22.2.27
81	Gnielka	Kurt	Mechanic	5.2.26	17.9.27	22.11.26
82	Grundmann	Alfred	Smith	5.2.26	10.3.26	———
83	Gerstaedt	Otto	Carpenter Foreman	24.3.26	21.5.28	26.3.27
84	Harbrecht	Willi	Electrical Fitter	6.1.26		22.11.26
85	Hehle	Leopold	Electrical Fitter	25.10.25	15.8.29	1.11.26
86	Heim	Georg	Motor Driver	20.9.25	2.10.27	12.7.27
87	Herzberg	Albert	Carpenter	1.11.25	27.9.26	10.12.26
88	Herst	Ernst	Machine Erector	22.1.26	3.9.26	———
89	Hewe	Guenther	Motor Driver	5.10.25	26.6.26	———
90	Huebel	Ernst	Carpenter	1.11.25	31.7.26	——-
91	Huber	Albert	Carpenter	1.11.25	23.10.27	No time condition
92	Huss	Friedrich	Carpenter Foreman	20.11.25		20.11.26
93	Hahn	Walter	Engine Driver	2.2.26	15.6.26	——
94	Hielscher	Alexander	Mechanic	2.2.26	24.5.29	22.11.26
95	Hoffmann	Michael	Foreman	2.2.26	May 1927	22.5.27
96	Horm	Karl	Mechanic	2.2.26	28.3.28	22.11.26
97	Hasenstab	Richard	Dredger Driver	5.2.26	24.8.29	22.2.27
98	Hermann	Werner	Mechanic	7.2.26	28.3.28	22.11.26
99	Hinrichsen	Andreas	Machinist	5.2.26	16.12.26	22.11.26
100	Hamel	Gustav	Machinery Foreman	1.3.26	9.7.26	——-

No:	Surname	Christian Name	Occupation	Date of Arrival	Date of Return	Existing Permit Expires
101	Heilmeier	Konrad	Engine Fitter	1.3.26	28.3.28	22.11.26
102	Hentschel	Otto	Engine Fitter	1.3.26		22.11.26
103	Holle	Walter	Machinist	1.3.26		22.11.26
104	Haas	Edmund	Smith	16.3.26		10.12.26
105	Hauck	August	Dredger Foreman	16.3.26	21.4.29	10.12.26
106	Hinze	Karl	Machinery Foreman	16.3.26	31.7.26	——-
107	Hoppe	Joseph	Machinery Foreman	24.3.26	19.10.26	26.3.27
108	Jonas	Albert	Foreman	5.10.25	18.2.19	12.1.27
109	Juhre	Karl	Electrical Foreman	29.11.25		20.1.27
110	Jochum	Marcellus	Dredger Foreman	27.2.26	28.3.28	26.3.28
111	Kachel	Konstantien	Carpenter	1.11.25	27.9.26	1.11.26
112	Kaestner	Karl	Dredger Foreman	6.1.26	15.6.26	——-
113	Kammerbauer	Johann	Foreman			
114	Karg	Johann	Concrete Specialist	13.12.25	16.9.28	22.11.26
115	Kasper	Alfred	Electrical Fitter	5.10.25	19.12.27	13.1.27
116	Keck	Joseph	Engine Driver	6.1.26	23.6.29	22.2.27
117	Kehl	Anton	Dredger Foreman	4.1.26	30.4.26	——-
118	Keller	Horst	Engine Fitter	21.9.26		12.1.27

– 4 –

No:	Surname	Christian Name	Occupation	Date of Arrival	Date of Return	Existing Permit Expires
119	Keller	Gustav	Foreman	20.11.25	4.10.25	10.12.26
120	Kiefer	Karl	Machine Fitter	10.12.25	24.8.29	10.12.26
121	Klein	Heinrich	Carpenter	15.1.26	27.9.26	22.5.27
122	Klein	Rudolf	Carpenter	15.1.26	27.9.26	22.5.27
123	Kleinbach	Friedrich	Mechanic	6.1.26	31.3.27	30.10.26
124	Klems	Peter	Machine Erector	6.1.26	3.10.26	15.10.26
125	Knappick	Paul	Dredger Foreman	6.1.26	9.5.27	No time condition
126	Knick	Franz	Machine Erector	15.1.26	26.6.26	——-
127	Kocher	Friedrich	Mechanic	21.9.26		12.1.27
128	Kochler	August	Erector for Iron construction	13.12.25	30.8.28	22.5.27
129	Kosboth	Willi	Motor Driver	9.1.26	30.4.26	——-
130	Koy	Karl	Carpenter	15.1.26	22.3.29	22.5.27
131	Krebs	Gustav	Machine Erector	15.1.26	26.6.26	——-
132	Krebs	Karl	Mechanic	21.9.25	24.8.29	12.1.27
133	Krieger	Linus	Machinist	1.11.25	31.10.29	1.11.26
134	Kriete	Johann	Mechanic	1.11.25		1.12.26
135	Kuch	Gottlieb	Mechanic	6.1.26	27.7.29	22.2.27

No:	Surname	Christian Name	Occupation	Date of Arrival	Date of Return	Existing Permit Expires
136	Kuhn	Joseph	Mechanic	21.9.25	16.4.28	12.1.27
137	Kummetz	Max	Machine Erector	22.11.25	5.3.26	——-
138	Kunz	Jakob	Foreman	10.12.25	†21.12.28	10.12.26
139	Kunz	Johann	Foreman	20.9.25	14.3.29	12.1.27
140	Kersten	Karl	Motor Fitter	2.2.26	23.6.29	No time condition
141	Kuhn	Karl	Mechanic	2.2.26		22.11.26
142	Klipp	Hermann	Machine Erector	12.2.26	26.6.26	——-
143	Kunert	Richard	Machine Erector	12.2.26	26.6.26	——-
144	Kramp	Richard	Foreman	1.3.26	4.10.29	22.5.27
145	Kormann	Joseph	Mechanic	26.2.26	12.4.26	——-
146	Lacker	Friedrich	Mechanic	21.9.25		12.1.27
147	Lessing	Max	Mechanic	1.11.25		1.11.26
148	Leu	Ernst	Engine Fitter	10.12.25	10.5.29	9.12.26
149	v. d. Lieth	Hermann	Engine Driver	13.12.25		22.2.27
150	Lintner	Hans	Machine Erector	15.1.26	9.7.26	——-
151	Luger	Joseph	Electrical Fitter	15.11.25		1.11.26
152	Luther	Fritz	Foreman for Drilling	13.12.25		22.11.26
153	Lutz	Otto	Mechanic	15.1.26	2.1.29	22.11.26
154	Louis	Joseph	Motor Fitter	26.12.25	23.11.27	No time condition
155	Linnenbrink	Ferdinand	Machinist	2.2.26	2.6.26	——-
156	Luedkehaus	Emil	Machine Erector	2.2.26		15.10.26
157	Lamprecht	Karl	Machinist	5.2.26	31.5.26	——-

– 5 –

No:	Surname	Christian Name	Occupation	Date of Arrival	Date of Return	Existing Permit Expires
158	Liese	Karl	Mechanic	5.2.26	24.8.29	22.11.26
159	Lieberam	Max	Foreman	1.3.26	22.5.26	——-
160	Ludwig	Alfred	Dredger Foreman	16.3.26	25.4.29	10.12.26
161	Luginger	Joseph	Dredger Foreman	24.3.26	27.7.29	26.3.27
162	Mathiesen	Heinrich	Carpenter Foreman	1.11.25		1.11.26
163	Maurer	Joseph	Foreman	10.12.25	17.1.27	——-
164	Meide	Bernhard	Carpenter	5.10.25	27.9.26	13.1.26
165	Mende	Arthur	Mechanic	1.11.25	23.10.27	1.11.26
166	Menzel	Franz	Foreman	22.1.26	5.3.26	——-
167	Mockert	Valentin	Mechanic	21.9.25		12.1.27
168	Mojzis	Wilhelm	Smith	21.9.25		12.1.27
169	Mueller	Alexander	Mechanic	1.11.25	British subject by birth	22.12.26

170	Mueller	Karl	Mechanic	15.1.26	11.4.26	
171	Mueller	Stephen	Concrete Specialist	22.1.26	16.12.28	23.1.27
172	Mueller	Bernhard	Mechanic	5.2.26	9.7.26	
173	Mueller	Georg	Mechanic	5.2.26		22.11.26
174	Mueller	Wolfgang	Machine Erector	23.2.26	27.9.26	30.9.26
175	Mueller	Adolf	Dredger Foreman	24.3.26	23.6.29	26.3.27
176	Muendoerfer	Anton	Mechanic	1.11.25	16.12.28	
177	Mundt	Ernst	Mechanic	21.9.25	24.8.29	12.1.27
178	Marciniak	Otto	Mechanic	5.2.26		22.11.26
179	Magel	Otto	Machine Erector	9.2.26	26.6.26	——-
180	Maschmann	Paul	Foreman	16.3.26	22.3.29	10.12.26
181	Meyer	Gottlieb	Dredger Foreman	16.3.26	11.5.28	
182	Nevir	Henry	Machinist	6.1.26		22.11.26
183	Nixdorf	Hermann	Machinist	5.10.25		14.1.27
184	Noe	Valentin	Electrical Fitter	1.11.25	23.6.29	22.11.26
185	Neumann	Karl	Belt Saddler	1.3.26	16.12.26	22.5.27
186	Oldenburg	Johannes	Machine Erector	18.11.25	22.6.26	
187	Otto	Max	Foreman	15.1.26	5.3.26	
188	Pacschik	Walter	Carpenter	1.11.25	27.9.26	
189	Paul	Johann	Joiner	1.11.25	27.9.26	No time condition
190	Pacuser	Paul	Joiner	1.11.25	14.10.27	No time condition
191	Pantzien	Otto	Machine Erector	10.12.25	22.6.26	——-
192	Plaesterer	Jakob	Electrical Fitter	22.1.26	24.5.29	22.2.27
193	Plickert	Wilhelm	Carpenter	1.11.25	8.8.26	——-
194	Del Puppo	Giovanni	Dredger Foreman	6.1.26	24.5.29	No time condition
195	Petermann	Gottfried	Mechanic	5.2.26	27.9.26	22.11.26
196	Pigors	Hermann	Concrete Specialist	5.2.26	26.3.28	22.5.27

– 6 –

No:	Surname	Christian Name	Occupation	Date of Arrival	Date of Return	Existing Permit Expires
197	Foetter	Paul	Foreman	7.2.26	7.2.28	22.5.27
198	Pellentzke	Fritz	Machine Erector	21.2.26	26.6.26	——-
199	Pichler	Leander	Dredger Foreman	1.3.26	12.4.29	22.5.27
200	Quick	Otto	Dredger Foreman	24.3.26	27.4.28	26.3.27
201	Raus	Karl	Mechanic	1.11.25	15.9.29	1.11.26
202	Rehbein	Hermann	Boiler Smith	22.1.26	5.4.28	22.11.26
203	Reuter	Karl	Smith	5.10.25	22.7.28	13.1.27
204	Reinschmidt	Franz	Explosive Foreman	13.12.25	16.9.28	13.12.26

205	Riemer	August	Electrical Fitter	6.1.26	27.7.29	22.11.26
206	Rosenfelder	Emil	Electrical Fitter	1.11.25		22.11.26
207	Roth	Leopold	Mechanic	1.11.25		9.12.26
208	Ruff	August	Foreman	10.12.25	24.12.26	1.12.26
209	Rattay	Max	Mechanic	2.2.26	29.9.26	22.11.26
210	Rieger	Wilhelm	Mechanic	2.2.26		22.11.26
211	Russ	Konrad	Electrical Erector	27.2.26	5.5.26	
212	Reissig	Kalr	Store Administrator	10.4.26		9.4.27
213	Samolewicz	Bruno	Machinery Foreman	4.10.25	22.9.26	No time condition
214	Seeger	Karl	Carpenter	1.11.25	24.5.29	No time condition
215	Suchowsky	Paul	Mechanic	21.9.25	19.12.27	12.1.27
216	Seberich	Leo	Foreman	2.2.26		22.5.27
217	Sauer	Philipp	Machinery Foreman	1.3.26	15.10.26	22.5.27
218	Siebert	Albert	Explosive Foreman	1.3.26	18.1.27	22.5.27
219	Sarcyka	Gustav	Foreman	24.3.26	24.5.29	263.27
220	Schaar	Fritz	Foreman	11.9.25	13.7.27	No time condition
221	Schaber	Fritz	Electrical Fitter	21.9.25	17.9.27	12.1.27
222	Scherbel	Anton	Smith	22.1.26	21.1.28	22.11.26
223	Schiel	Franz	Mechanic	21.9.25	9.8.26 & 23.4.29	12.1.27
224	Schillinger	Franz	Machinery Foreman	21.9.25	20.7.29	12.1.27
225	Schmidt	Erich	Mechanic	21.9.25	20.2.26	———
226	Schnaerz	Otto	Dredger Foreman	5.10.25		13.1.27
227	Schneider	Ernst	Mechanic	1.11.25		No time condition
228	Schneider	Ernst	Mechanic & Painter	15.1.26	11.1.28	10.12.26
229	Schoerger	Ludwig	Electrical Foreman	22.1.26	24.5.29	22.11.26
230	Schuhmacher	Johann	Mechanic	1.11.25	6.4.28	No time condition
231	Schulz	Hermann	Concrete Specialist	6.1.26	3.3.28	No time condition
232	Schaeffer	Willi	Machinist	5.2.26	22.2.29	22.2.27
233	Schell	Joseph	Machinist	5.2.26	9.6.27	22.11.26
234	Schulze	Johannes	Machine Erector	23.3.26	29.7.27	30.9.26
235	Schramm	Paul	Carpenter Foreman	1.3.26	28.3.28	22.5.27
236	Stache	August	Dredger Foreman	6.1.26	21.9.28	No time condition

– 7 –

No:	Surname	Christian Name	Occupation	Date of Arrival	Date of Return	Existing Permit Expires
237	Stahl	Johannes	Machinery Foreman	6.1.26		22.11.26

238	Stephan	Ludwig	Electrical Fitter	1.11.25	4.10.29	22.11.26
239	Sterz	Karl	Carpenter	15.1.25	27.9.26	22.5.27
240	Stell	August	Mechanic	6.1.26	19.9.26	22.11.26
241	Stuewe	Hans	Carpenter	1.11.25	27.9.26	22.5.27
242	Stoeckl	Franz	Concrete Specialist	2.2.26		2.2.27
243	Strauch	Hans	Mechanic	5.2.26	28.11.26	22.11.26
244	Steinemann	Alphons	Machine Erector	12.2.26	26.6.26	——-
245	Strehle	Emil	Electrical Fitter	1.3.26	27.7.29	22.11.26
246	Thaeder	Wilhelm	Erector	3.12.25	15.5.26	——-
247	Theiss	Nikolaus	Dredger Foreman	15.11.25	24.5.29	22.5.27
248	Theiss	Ludwig	Dredger Foreman	22.1.26	24.12.26	22.5.27
249	Tjardes	Richard	Foreman	11.9.25	27.6.28	No time condition
250	Tuebbecke	Paul	Foreman	18.11.25		17.11.26
251	Trost	Friedrich	Mechanic	5.2.26		22.11.26
252	Uherek	Franz	Machine Erector	17.4.26	17.9.26	
253	Voges	Kurt	Dredger Foreman	5.10.25		12.1.27
254	Volz	Alois	Mechanic	21.9.25	24.8.29	12.1.27
255	De Vries	Albertus	Electrical Foreman	5.10.25	4.10.29	12.1.27
256	Vess	Paul	Concrete Specialist	2.2.26	21.3.26	——-
257	Veller	Hubert	Mechanic	7.2.26	7.9.28	22.11.26
258	Vandersee	Otto	Mechanic	12.2.26		22.11.26
259	Waag	Otto	Mechanic	1.11.25	7.9.28	1.11.26
260	Wagner	Joseph	Machinist	2.11.25	29.7.29	1.11.26
261	Wagner	Paul	Mechanic	15.1.26		22.11.26
262	Wahl	August	Mechanic	21.9.25	7.1.28	12.1.27
263	Walter	Franz	Mechanic	10.12.25		9.12.26
264	Wassenberg	Friedrich	Mechanic	23.9.25	29.1.27	12.1.27
265	Weise	Franz	Erector f.Ironconstr.	22.1.26		30.10.26
266	Wenzel	Eduard	Mechanic	1.11.25	25.5.28	11.11.26
267	Werner	Heinrich	Mechanic	15.1.26	24.8.29	22.11.26
268	Wiegele	Franz	Machinist	1.11.25	27.7.29	1.11.26
269	Wiemann	Friedrich	Machine Erector	15.1.26	24.3.26	
270	Wienecke	Wilhelm	Carpenter	1.11.25	27.9.28	
271	Wilhelmi	Walter	Explosive's Foreman	15.1.26	4.8.29	No time condition
272	Winterer	Karl	Mechanic	1.11.25	15.6.26	
273	Wittfoth	Karl	Carpenter	1.11.25	27.9.26	No time condition
274	Woelfle	Julius	Mechanic	21.9.25	27.4.28	12.1.27
275	Wunsch	Franz	Foreman	11.9.25	24.8.29	12.1.27
276	Wustrach	Gustav	Welder	23.1.26		22.11.26

No:	Surname	Christian Name	Occupation	Date of Arrival	Date of Return	Existing Permit Expires
277	Wiedmaier	Georg	Concrete Specialist	2.2.26	24.8.29	
278	Wilmers	Wilhelm	Engine Driver	2.2.26	8.8.26	
279	Wiehert	Kurt	Mechanic	2.2.26	11.4.27	22.11.26
280	Weichbrod	Georg	Electrical Fitter	5.2.26	23/4/29	22.11.26
281	Weckebrodt	Eduard	Machine Erector	12.2.26	9.9.26	
282	Weisser	Christian	Smith	16.3.26		10.12.26
283	Wimmer	Matthias	Dredger Foreman	16.3.26	24.5.29	10.12.26
284	Walter	Heinrich	Dredger Foreman	24.3.26		26.3.27
285	Weber	Gustav	Dredger Foreman	24.3.26		26.3.27
286	Winkler	Georg	Foreman	24.3.26	27.7.29	26.3.27
287	Wollenschlaeger	Albert	Foreman	24.3.26	4.11.26	26.3.27
288	Zaiss	Heinrich	Mechanic	1.11.25	10.2.29	1.11.26
289	Ziegler	Karl	Mechanic	21.9.25	29.10.27	13.1.26
290	Zuefle	Ernst (1)	Explosives Specialist	13.12.25	25.4.29	13.12.26
291	Zuefle	Ernst (2)	Explosives Specialist	13.12.25	25.4.29	13.12.26
292	Zutavern	Heinrich	Mechanic	21.1.25		12.1.27
293	Frohberg	Friedrich	Machine Erector	17.4.26	27.9.26	17.10.26
294	Hechne	Wilhelm	Machine Erector	25.4.26	26.6.26	
295	Kulemann	Oscar	Machine Erector	20.3.26	26.6.26	
296	Liebscher	Oswald	Machine Erector	17.4.26	26.6.26	
297	Otto	Ernst	Machine Erector	11.5.26	26.6.26	
298	Pepp	Bruno	Dredger Foreman	11.5.26	27.7.29	11.5.27
299	Reifergerste	Friedrich	Dredger Foreman	11.5.26	31.10.29	11.5.27
300	Rottweiler	Erich	Concrete Specialist	6.1.26	22.6.28	No time condition
301	Neumann	Erich	Erector	15.5.26	9.9.26	
302	Stuctzer	Walter	Erector	25.5.26	9.6.26	
303	Henney	Karl	Student	29.5.26	30.9.26	
304	Dietrich	Wilhelm	Machine Erector	29.5.26	21.7.26	
305	Andres	Willi	Dredger Foreman	25.6.26	12.7.29	4.6.27
306	Barlau	Walter	Machinist	5.6.26	27.7.29	4.12.26
307	Bauer	Lorenz	Dredger Foreman	5.6.26	24.5.29	4.6.27
308	Behrens	Martin	Electrical Fitter	5.6.26		4.6.27
309	Burkhardt	Gottfried	Dredger Foreman	5.6.26	7.6.29	4.6.27
310	Gernsbeck	Leonhard	Electrical Fitter	5.6.26	Died 9.7.26	
311	Guenther	Adam	Dredger Foreman	5.6.26	30.8.28	4.6.27
312	Haas	Franz	Dredger Foreman	5.6.26	8.7.28	4.12.27

313	Haas	Otto	Dredger Foreman	5.6.26	8.7.28	4.6.27
314	Hock	Julian	Dredger Foreman	5.6.26	24.8.29	4.6.27
315	Knosalla	Albert	Dredger Foreman	5.6.26	24.8.29	
316	Kuehr	Friedrich	Diesel Engine Driver	5.6.26		

– 9 –

No:	Surname	Christian Name	Occupation	Date of Arrival	Date of Return	Existing Permit Expires
317	Mittank	Karl	Dredger Foreman	5.6.26	4.10.29	
318	Pankratz	Joseph	Engine Driver	5.6.26	22.6.28	4.6.27
319	Reifegerste	Wilhelm	Dredger Foreman	5.6.26	2.11.28	4.6.27
320	Reuther	Karl	Dredger Foreman	5.6.26	30.11.28	4.6.27
321	Schreiber	Emil	Dredger Foreman	5.6.26		4.6.27
322	Wensauer	Johann	Dredger Foreman	5.6.26	16.8.26	——-
323	Zimmerling	Oskar	Dredger Foreman	5.6.26		4.12.26
324	Heinze	Ernst	Machine Erector	8.6.27	8.2.27	8.6.27
325	Ruthmann	Karl	Mechanic	21.9.25	4.10.25	——-
326	Preise	Friedrich	Mechanic	21.9.25	4.10.25	——-
327	Sfimmig	Reinhold	Mechanic	21.9.25	15.12.25	——-
328	Gantzer	Karl	Mechanic	12.6.26		12.12.26
329	Krause	Otto	Diesel Erector	12.6.26	14.3.28	
330	Ebernhard	Ernst	Foreman	12.6.26	21.7.27	12.12.26
331	Puffer	Johann	Foreman	12.6.26		12.12.26
332	Stockinger	Johann	Foreman	12.6.26	4.8.29	12.12.26
333	Bernold	Hans	Student (Apprentice)		8.6.28	
334	Wienecke	Erich	Student (Apprentice)	29.5.26	27.9.26	
335	Arnitz	Alois	Dredger Foreman	23.6.26	16.8.26	——-
336	Warchowka	Georg	Iron Nettings Specialist	2.7.26	27.9.26	2.1.27
337	Geyer	Adolf	Iron Nettings Specialist	2.7.26	7.6.28	2.1.27
338	Korona	Karl	Iron Nettings Specialist	2.7.26	16.12.26	2.1.27
339	Schacht	Anton	Diver	6.7.26	15.1.29	——-
340	——-					
341	Wahl	Theodor	Apprentice			
342	Hock	Theodor	Foreman	24.7.26	6.4.27	No time condition
343	Fuerst	Joseph	Foreman	24.7.26	27.7.29	
344	Sonnderfer	Joseph	Foreman	24.7.26	6.1.28	
345	Walter	Gustav	Foreman	24.7.26	7.6.29	
346	Waibel	Hermann	Foreman	26.7.26		26.1.27
347	Schopf	Gottfried	Machine Erector	26.7.26	6.1.28	26.10.27
348	Preuse	Fritz	Student	28.7.26	30.9.26	

349	Liedtke	Hellmuth	Student	6.8.26	19.10.26	
350	Schiffbaenker	Johann	Foreman	9.8.26	24.5.29	
351	Luginger	Wilhelm	Dredger Foreman	13.8.26	27.8.29	
352	Albert	Andreas	Foreman	13.8.26	23.6.29	
353	Spanjer	Meinhard	Foreman	21.8.26	27.7.29	20.2.27

– 10 –

No:	Surname	Christian Name	Occupation	Date of Arrival	Date of Return	Existing Permit Expires
354	Schroeder	Egon	Foreman	21.8.26	20.3.27	20.2.27
355	Bronold	Martin	Engine Driver	21.8.26	26.9.27	20.2.27
356	Degorski	Hermann	Engine Driver	21.8.26	30.8.29	20.2.27
357	Grucza	Anton	Engine Driver	21.8.26	15.10.29	20.2.27
358	Schulz	Gustav	Engine Driver	21.8.26	11.11.28	20.2.27
359	Harlander	Otto	Engine Driver	21.8.26	31.10.29	20.2.27
360	Theiss	Joseph	Engine Driver	21.8.26	16.4.28	20.2.27
361	Sindermann	Joseph	Foreman f. Drilling	21.8.26	10.5.29	20.2.27
362	Mueller	Karl	Engine Driver	21.8.26	22.3.29	20.2.27
363	Harlander	Karl	Engine Driver	21.8.26	3.8.27	20.2.27
364	Nicske	Hugo	Engine Driver	21.8.26	28.9.26	20.2.27
365	Mayr	Maximilian	Engine Driver	21.8.26		20.2.27
366	Kaeshammer	Franz	Machinist	21.8.26	23.6.29	20.2.27
367	Kummerow	Hans	Student	21.8.26		2.10.26
368	Kress	Kurt	Student	24.8.26	6.10.26	
369	Gruening	Guenther	Student	23.8.26	20.9.26	
370	Guterunst	Friedrich	Machinery Foreman	27.8.26	11.12.28	26.2.27
371	Neubauer	Wilhelm	Dredger Foreman	27.8.26	31.10.27	26.2.27
372	Schulz	Wilhelm	Engine Driver	27.8.26	24.5.29	26.2.27
373	Schwensow	Paul	Copper Smith	27.8.26	30.8.28	26.2.27
374	Klumpp	Gustav	Mechanic	4.9.26	4.10.29	4.3.27
375	Hussy	Willibald	Dredger Foreman	4.9.26	8.8.28	4.3.27
376	Vierthaler	Georg	Dredger Foreman	4.9.26		4.3.27
377	Werner	Michael	Foreman	4.9.26		4.3.27
378	Mattern	Hanz	Student	2.10.26	24.9.26	
379	Vollmer	Wilhelm	Cook	4.10.26	12.4.28	
380	Erdmann	Reinhold	Carpenter	4.10.26	22.3.29	
381	v. Czapiewski	Waldemar	Joiner	4.10.26	10.12.27	
382	Reinicke	Otto	Joiner	4.10.26	23.6.29	
383	Fischer	Otto	Carpenter	4.10.26	23.6.29	
384	Lulei	Fritz	Carpenter	6.10.26	18.11.26	

385	Milberg	Max	Carpenter	6.10.26	22.3.29	
386	Weilandt	Christian	Carpenter	6.10.26		
387	Wilde	Paul	Carpenter	6.10.26	31.10.29	
388	Sohner	Fritz	Carpenter	6.10.26	31.10.29	
389	——-	——-	——-	——-		
390	——-	——-	——-	——-		
391	Arnhold	Fritz	Fitter	13.11.26	21.1.28	
392	Englert	Willibald	Digger Driver	13.11.26		
393	Marienfeld	Albert	Fitter	13.11.26	20.3.27	

– 11 –

No:	Surname	Christian Name	Occupation	Date of Arrival	Date of Return	Existing Permit Expires
394	Paesch	Wilhelm	Fitter	13.11.26	25.5.27	
395	Ross	Josef	Machinist	13.11.26	27.7.29	
396	Schulz	Walter	Fitter	13.11.26	18.11.26	
397	Utech	Max	Foreman	13.11.26	29.2.28	
398	Zeller	Karl	Borer Grinder	13.11.26	7.3.28	
399	Harr	Gottlob	Explosive Specialist	19.11.26	24.8.29	
400	Walz	Christian	Explosive Specialist	19.11.26	30.6.29	
401	Merkel	Ludwig	Explosive Specialist	19.11.26	9.8.29	
402	Gretz	Eugen	Explosive Specialist	19.11.26	24.8.29	
403	Guhl	Jakob	Welder	29.11.26		
404	Haube	Richard	Rope Maker	14.12.26	28.2.27	
405	Hann	Anton	Excavator Foreman	14.12.26	6.8.27	
406	Burtscher	Gelehard	Cable Crane Fitter	14.12.26	24.5.29	
407	Petermann	Gottfried	Mechanic	14.12.26	31.1.27	
408	Stuewe	Heinrich	Carpenter	14.12.26	4.10.29	
409	Gauss	Friedrich	Foreman for Drilling	14.12.26	20.7.29	
410	Streich	Heinrich	Dredger Foreman	14.12.26	7.6.29	
411	Dierkopf	Max	Carpenter	14.12.26	4.10.29	
412	Klein	Heinrich	Carpenter	14.12.26	22.3.29	
413	Lou	Karl	Specialist for Diesel engine on motor boat	16.12.26		
414	Hille	Hermann	Fitter	8.1.27	21.9.28	
415	Kelm	Otto	Special Fitter for stone crusher and washing plant	29.1.27		
416	Horst	Ernst	Dredger Foreman	29.1.27	31.6.29	
417	Kuenkel	Franz	Machinist	29.1.27	7.10.27	
418	Jahn	August		29.1.27	7.8.29	
419	Werner	Walter	Special Fitters for stone	29.1.27		

No:	Surname	Christian Name	Occupation	Date of Arrival	Date of Return	Existing Permit Expires
420	Kubicak	Paul	Crusher and washing plant	29.1.27	17.9.27	
421	Hiltmann	Robert		29.1.27		
422	Kuenzli	Albert	Dredger Foreman	29.1.27	11.3.27	
423	Schulz	Wilhelm	Special Fitter for bucket dredgers	8.2.27		
424	Scheuer	Albert	Special Fitter for locomotives	8.2.27		
425	Schueck	August	Special Fitter for stone crusher & washing plant	8.2.27	19.-.29	
426	Rutka	Richard	Special Fitter for Bucket Dredgers	8.2.27	12.3.28	
427	Heyer	Alwin	Special Fitters for locomotives	8.2.27	4.10.29	
428	Hilliger	Otto	Special Fitter for locomotives	8.2.27	12.7.29	
429	Hussy	Ambros	Dredger Driver	21.2.27	15.8.27	
430	Schlee	Georg	Rock-Driller	21.2.27	19.2.29	

– 12 –

No:	Surname	Christian Name	Occupation	Date of Arrival	Date of Return	Existing Permit Expires
431	Bieber	Georg	Fitter for stone crusher & washing plant	21.2.27	7.5.27	
432	Herrmann	Anton	Fitter for Bucket Dredgers	21.2.27	4.10.29	
433	Roth	Karl	Rock-Driller	21.2.27	2.11.28	
434	Huettinger	Joseph	Locomotive-Fitter	21.2.27		
435	Heitberger	Johann	Locomotive-Fitter	21.2.27	10.5.29	
436	Ziegler	Karl	Rock-Driller	26.2.27	12.4.29	
437	Weigold	Gottlieb	Rock-Driller	26.2.27	12.4.29	
438	Doelker	Mathaeus	Rock-Driller	26.2.27	12.4.29	
439	Buch	Albert	Rock-Driller	26.2.27		
440	Schuck	Josef	Engine Driver	6.3.27	7.5.27	
441	Schoenstetter	Josef	Engine Driver	6.3.27	12.7.29	
442	Reisinger	Josef	Engine Driver	6.3.27	12.7.29	
443	Birnkammer	Anton	Engine Driver	6.3.27		
444	Greiser	Paul	Engine Driver	6.3.27	12.7.29	
445	Lyginger	Franz	Engine Driver	6.3.27	12.11.28	
446	Lang	Ludwig	Engine Driver	6.3.27	12.7.29	
447	Hock	Egid	Engine Driver	6.3.27	24.5.29	
448	Lebermeyer	Karl	Engine Driver	6.3.27	12.4.29	
449	Schueler	Wilhelm	Excavator Foreman	6.3.27	?22/3/29	
450	Eichinger	Georg	Excavator Foreman	6.3.27	25.11.27	
451	Vogel	Willy	Engine Driver	6.3.27	7.9.28	

452	Psurek	Josef	Excavator Foreman	6.3.27	27.7.29	
453	Buettner	Matthias	Dredger Foreman	6.3.27	26.11.27	
454	Riemer	Karl	Engine Driver	6.3.327	30.6.29	
455	Handschaek	Friedrich	Engine Driver	6.3.27	16.1.29	
456	Liebherr	Ludwig	Mechanic	6.3.27	10.5.29	
457	Diebold	Otto	Rockdriller	6.3.27		
458	Piwonski	Karl	Specialist for making iron nettings	6.3.27	8.6.28	
459	Heer	Hermann	Engine Driver	6.3.27		
460	Branitzki	Erwin	Special Fitter for stone crusher & washing plant	21.3.27		
461	Schulz	Emil	Rockdriller	21.3.27	15.12.27	
462	Roesner	Karl	Excavator Foreman	21.3.27	24.5.27	
463	Drysalla	Wilhelm	Engine Driver	21.3.27		
464	Waag	Ludwig	Engine Driver	21.3.27	9.8.29	
465	Richter	Kurt	Cable crane Driver	21.3.27	24.5.29	
466	Da Corta	Paul	Electric loco Driver	24.3.27	22.7.27	
467	Klein	Rudolf	Carpenter	24.3.27	4.10.29	
468	Engel	Paul	Electric loco Driver	24.3.27	18.5.27	

– 13 –

No:	Surname	Christian Name	Occupation	Date of Arrival	Date of Return	Existing Permit Expires
469	Sterz	Karl	Carpenter	24.3.27		
470	Sinowsky	Alfred	Fitter for Belgian Rock drilling plant	24.3.27	15.9.29	
471	Stitzl	Johann	Dredger Foreman	24.3.27	?(22/3)-29	
472	Huber	Josef	Electric loco Driver	24.3.27	22.3.29	
473	Andrae	Rudolf	Electric loco Driver	24.3.27		
474	Meyer	Rudolf	Swivel crane operator	24.3.27	28.3.28	
475	Gebhardt	Paul	Fitter for cable crane	244.3.27	24.5.29	
476	Crusca	Paul	Electric loco Driver	24.3.27		
477	Englert	Johann	Excavator Foreman	24.3.27	4.8.29	
478	Nilpert	Johann	Excavator Foreman	26.3.27	23.6.27	
479	Niemand	Wilhelm	Excavator Foreman	26.3.27	25.6.27	
480	Horner	Josef	Excavator Foreman	26.3.27	†12.10.28	
481	Schaeffler	Martin	Excavator Foreman	26.3.27	25.4.29	
482	Schoepwinkel	Heinrich	Excavator Foreman	2.4.27	5.4.27	
483	Gunst	Joseph	Excavator Foreman	2.4.27	14.8.27	
484	Berndt	Paul	Excavator Foreman	2.4.27		
485	Voeckler	Albert	Excavator Foreman	2.4.27		
486	Weinlaeder	Georg	Excavator Foreman	2.4.27	15.9.29	

487	Kaltenecker	Georg	Excavator Foreman	2.4.27	21.5.27	
488	Mueller	Paul Otto	Excavator Foreman	2.4.27	20.4.29	
489	Santner	Alois	Excavator Foreman	2.4.27	4.10.29	
490	Gottsmann	Anton	Excavator Foreman	2.4.27	14.8.27	
491	Gluba	Max	Concrete Worker	12.4.27	14.10.27	
492	Gruebnau	Hermann	Concrete Worker	12.4.27	8.6.28	
493	Merz	Ernst	Engine Driver	12.4.27	24.8.29	
494	Tjardes	Karl	Excavator Foreman	13.5.27	15.2.28	
495	Wilke	Max	Swivel Crane Driver	13.5.27		
496	Schoenmetz	Karl	Reinforced Concrete Worker	13.5.27	7.7.29	
497	Karschau	Erich	Mechanic	13.5.27	4.10.29	
498	Klein	Bernhard	Swivel Crane Driver	13.5.27		
499	Grettke	Hans	Excavator Driver	13.5.27	24.8.29	
500	Droege	Wilhelm	Excavator Driver	13.5.27	2.10.27	
501	Schillinger	Otto	Mechanic	13.5.27	15.9.29	
502	Bartschicz	Paul	Mechanic	13.5.27	25.5.28	
503	Rolff	Kurt	Swivel Crane Fitter	13.5.27	27.7.28	
504	Montebaur	Georg	Fitter for stone-crushing & washing plant	13.5.27	3.10.28	
505	Huebner	Max	Diesel Engine Mechanic	13.5.27	25.5.28	
506	Trauschke	Kurt	Mechanic	13.5.27		
507	Luetkehaus	Paul	Mechanic	13.5.27		

No:	Surname	Christian Name	Occupation	Date of Arrival	Date of Return	Existing Permit Expires
508	Nock	Joseph	Mechanic	13.5.27	20.7.29	
509	Hock	Lorenz	Excavator Foreman	13.5.27	7.2.29	
510	Hogrefe	Wilhelm	Specialist for Electric Railways	13.5.27	10.7.27	
511	Pazderski	Joseph	Mechanic	13.5.27	7.6.29	
512	Gassner	Ludwig	Excavator Foreman	13.5.27	25.4.29	
513	Gutekunst	Karl	Mechanic	13.5.27	10.5.29	
514	Kundshagen	Heinrich	Winch Foreman	13.5.27	15.8.29	
515	Luz	Christian	Mechanic	13.5.27	10.5.29	
516	Link	Paul	Swivel Crane Driver	30.5.27		
517	Nayer	Johann	Carpenter	30.5.27	21.3.29	
518	Frick	Herrmann	Carpenter	30.5.27	31.1.29	
519	Junker	Wendelin	Mechanic	30.5.27	19.4.28	
520	Wagenlehner	Anton	Excavation Foreman	7.6.27	4.10.29	
521	Paul	Anton	Excavation Foreman	7.6.27	31.8.29	

No:	Surname	Christian Name	Occupation	Date of Arrival	Date of Return	Existing Permit Expires
522	Hoyer	Richard	Excavation Foreman	7.6.27	28.3.28	
523	Englert	Karl	Excavator Driver	15.6.27	24.8.29	
524	Hann	Robert	Excavator Driver	15.6.27		
525	Bonauer	Alois	Excavation Foreman	15.6.27	27.7.29	
526	Winkel	Wilhelm	Carpenter	15.6.27		
527	Zutavern	Wilhelm	Loco-Driver	15.6.27		
528	Maida	Paul	Excavation Foreman	18.6.27	23.6.29	
529	Graeber	Hubert	Student	18.6.27	30.7.27	
530	Berlin	Paul	Excavation Foreman	25.6.27	@8.7.27	
531	Plapperer	Jacob	Excavation Foreman	7.7.27		
532	Zaun	Fritz	Excavation Foreman	7.7.27	20.12.28	
533	Rippberger	Philipp	Excavation Foreman	7.7.27	27.7.29	
534	Groenger	Wilhelm	Excavation Foreman	7.7.27	23.6.29	
535	Berger	Ludwig	Excavation Foreman	7.7.27		
536	Lou	Anton	Excavation Foreman	7.7.27		
537	Richter	Hermann	Dredger Foreman	9.7.27	12.7.29	
538	Schentzel	Emil	Mechanic	9.7.27		
539	Rolin	Heinrich	Mechanic	9.7.27	11.11.28	
540	Pelloth	Fritz	Excavation Foreman	9.7.27	27.7.29	
541	Nagdeburg	Otto	Mechanic	9.7.27	25.11.27	
542	Nuny	Germann	Student	9.7.27	20.7.27	
543	Hemmerich	Josef	Excavation Foreman	16.7.27	27.9.27	
544	Hussy	Hermann	Electric Loco Driver	16.7.27	10.5.29	
545	Ebert	Joseph	Dredger Foreman	30.7.27	22.11.27	
546	Smolla	Karl Wilhelm	Student	30.7.27	20.10.27	

– 15 –

No:	Surname	Christian Name	Occupation	Date of Arrival	Date of Return	Existing Permit Expires
547	Lutz	Fritz	Student	30.7.27	23.10.27	
548	Fuhrmann	Otto	Student	4.8.27	23.10.27	
549	Schulte	Terris	Mechanic	8.8.27	10.9.28	
550	Bayer	Anton	Excavation Foreman	8.8.27	23.6.29	
551	Stoerr	Albert	Student	8.8.27	23.10.27	
552	Schmidt	Johannes	Special Erector for Krupp-Diesel engines	28.6.27	1.11.27	
553	Huppekothem	Wilhelm	Electrician	12.8.27	20.7.29	
554	Schaffrath	Johannes	Excavation Foreman	12.8.27	17.5.29	
555	Bochert	Karl	RockDriller	12.8.27	14.9.27	
556	Neiminger	Martin	Mechanic	12.8.27	1.6.29	
557	Nozcheinen	Johannes	Excavation Foreman	12.8.27	24.8.29	

558	Hentschel	Johannes	Mechanic	12.8.27		
559	Bushler	Andreas	Rockdriller	12.8.27	15.8.29	
560	Kallenbach	Wilhelm	Student	12.8.27	7.10.27	
561	Laudien	Walter	Student	16.8.17	7.10.27	
562	Blekenbach	Helmuth	Student	20.8.27	23.10.27	
563	Nyeliwictz	Richard	Excavation Foreman	24.8.27	3.3.29	
564	Bauer	Ludwig	Carpenter	24.8.27	9.8.29	
565	Kayer	Stephan	Excavation Foreman	24.8.27	2.11.28	
566	Grabocski	Gerhard	Student		27.10.27	
567	Steffen	Hermann	Excavation Foreman		24.8.29	
568	Maier	Anton	Engine Driver			
569	Meshrke	Richard	Excavator Driver		9.8.29	
570	Noll	Karl	Mechanic			
571	Wittfoth	Karl	Carpenter for reinforced concrete			23.6.29
572	Rolff	Bruno	Special Fitter for stone crushing plant			22.6.28
573	Wichofsky	Conrad	Carpenter		24.8.29	
574	Spath	Johann	Excavation Foreman		24.5.29	
575	Artinger	Thomas	Carpenter		5.12.27	
576	Schmidt	Franz	Carpenter		22.2.29	
577	Reuss	Philipp	Locomotive Driver			
578	Schimpf	Franz	Excavator Driver		19.7.28	
579	Gysinski	Ludwig	Rockdriller		15.7.28	
580	Petong	Ernst	Mechanic			
581	Kuke	Ewald	Joiner		3.5.29	
582	Schuetz	Albert	Excavation Foreman		8.10.28	
583	Aulbach	Michael	Excavator Driver		15.1.29	
584	Lepsch	Ernst	Concrete Worker			
585	Aulbach	Rudolf	Excavator Driver		23.6.29	

– 16 –

No:	Surname	Christian Name	Occupation	Date of Arrival	Date of Return	Existing Permit Expires
586	Knoch	Arno	Carpenter		23.5.28	
587	Mehlkorn	Franz	Carpenter		22.3.29	
588	Soldmann	Max	Mechanic			
589	Pella	Joseph	Dredger Driver		23.4.29	
590	Wallsteiner	Erich	Fitter for Locomotives			
591	Lauder	Georg	Fitter for Locomotives			
592	Feerster	Franz	Dredger Driver		11.11.28	
593	Staab	Richard	Dredger Driver		30.6.29	

No	Surname	Christian Name	Occupation	Date of Arrival	Date of Return	Existing Permit Expires
594	Marr	Willy	Specialist for Cranes		7.7.29	
595	Riebl	Ludwig	Specialist for Electric Railways		16.9.28	
596	Stadlinger	Bartholomaus	Dredger Driver			
597	Loeffler	Max	Dredger Driver			
598	Brueckl	Georg	Rockdriller		12.7.29	
599	Zoecke	Erich	Cable-Crane Driver			
600	Luz	Wilhelm	Locomotive Fitter		30.11.28	
601	Wolters	Hermann	Mechanic			
602	Gaiser	Hermann	Granite Worker		6.8.28	
603	Braun	Arhur	Granite Worker		6.8.28	
604	Fischer	Martin	Rockdriller		12.7.29	
605	Guenther	Hermann	Granite Worker			
606	Pletz	Joseph	Mechanic		12.7.29	
607	Raila	Martin	Excavator Driver		7.6.29	
608	Schimonski	Bruno	Engine Fitter		22.5.28	
609	Taeuber	Christian	Granite Worker		27.2.29	
610	Vosz	Emil	Excavation Foreman			
611	Kapperer	Fritz	Granite Worker		27.8.28	
612	Cieluck	Stanislaus, Johannes	Excavation Foreman		4.10.29	
613	Fassarek	Kurt	Reinforced Concrete Worker		25.5.29	
614	Markstahler	Heinrich	Ecavation Foreman		27.7.29	
615	Schadt	Friedrich	Special Granite Worker		27.4.28	
616	Wirth	Franz	Special Granite Worker			
617	Zoz	Lorenz	Special Granite Worker		7.3.28	
618	Kerschner	Valentin	Excavation Foreman			
619	Kobler	Otto	Excavation Foreman		11.11.28	
620	Berger	Johann	Excavation Foreman		23.6.29	
621	Glaser	Gottlieb	Foreman for Rockdrilling		11.11.28	
622	Fischer	Gregor	Reinforced Concrete Worker			
623	Schoenmetz	Ansgar	Reinforced Concrete Worker			
624	Beilharz	Matthaeus	Rockdriller		12.4.29	
625	Guehl	Richard	Carpenter		28.3.28	

– 17 –

No:	Surname	Christian Name	Occupation	Date of Arrival	Date of Return	Existing Permit Expires
626	Schneider	Karl	Reinforced Concrete Worker			8.6.28
627	Wienecke	Wilhelm	Carpenter		28.3.28	
628	Stockinger	Franz	Excavation Foreman		23.6.29	

629	Heinen	Hermann	Special Mechanic for Diesel Engines		4.8.29	
630	Zellerhorst	Wilhelm	Special Mechanic for Diesel Engines			
631	Skrypczak	Anton	Excavation Foreman		24.5.29	
632	Dassek	Friedrich	Cook			
633	Franke	Willy	Mechanic		3.6.28	
634	Wendtorf	Hans	Mechanic			
635	Bestz	Christopher	Carpenter		19.7.28	
636	Friedrich	Arthur	Mechanic-Foreman		23.6.29	
637	Schlegel	Bernhard	Excavation Foreman		2.11.28	
638	Scheffler	Otto	Excavation Foreman		21.11.28	
639	Goldschmidt	Walter	Excavation Foreman		12.4.29	
640	Hanke	Alfred	Mechanic		24.8.29	
641	Raabe	Rudolf	Mechanic		3.4.29	
642	Nolden	Theodor	Mechanic		30.6.29	
643	Schoebel	Kurt	Dredger Driver		4.10.29	
644	Gernoth	Arthur	Mechanic			
645	Rhinow	Walter	Mechanic		7.6.29	
646	Schaefer	Fritz	Mechanic		12.10.28	
647	Kuehne	Hermann	Mechanic			
648	Prodlick	Franz	Mechanic			
649	Scheckenbach	Franz	Excavation Foreman		25.4.29	
650	Thaeder	Wilhelm	Carpenter			
651	Homburg	Gustav	Carpenter		22.2.29	
652	Kempken	Johann	Locomotive Driver		6.10.28	
653	Hullmann	Erwin	Mechanic			
654	Kohl	Wilhelm	Carpenter			
655	Breyer	Emil	Locomotive Driver		2.1.29	
656	Lacker	Hans	Mechanic		20.7.29	
657	Kabisch	Emil	Excavation Foreman		22.2.29	
658	Czichas	Urban	Excavation Foreman		27.7.29	
659	Mueller	Gustav	Locomotive Fitter		15.9.29	
660	Kaiser	Waldemar	Carpenter		26.10.28	
661	Bahre	Emil	Excavation Foreman		24.8.29	
662	Babilon	Johannes	Dredger Operator		13.2.28	
663	Picchulla	Karl	Locomotive Driver		19.5.29	

No:	Surname	Christian Name	Occupation	Date of Arrival	Date of Return	Existing Permit Expires
664	Bulkowski	Bernhard	Carpenter		26.10.28	
665	Smoczyk	Felix	Locomotive Driver		10.5.29	
666	Walter	Alexander	Excavation Foreman		23.6.29	
667	Raedel	Ernst	Mechanic			
668	Pasemann	Otto	Carpenter		26.10.28	
669	Schiffkowitz	Willi	Carpenter		26.10.28	
670	Zielinski	Anastasius	Dredger Operator		4.10.29	
671	Schoenebeck	Richard	Locomotive Fitter			
672	Engel	Franz	Excavation Foreman		22.2.29	
673	Karthaeuser	Johann	Carpenter		26.10.28	
674	Klier	Rudolf	Mechanic		11.11.28	
675	Beduarek	Josef	Mechanic		16.8.29	
676	Berucki	Thomas	Fitter for iron constructions		5.5.29	
677	Koserowski	Oskar	Excavation Foreman			
678	Brunold	Nicolaus	Dredger Operator		9.8.29	
679	Schimpf	Ludwig	Dredger Operator		25.4.29	
680	Stegmann	Ferdinand	Carpenter		15.9.29	
681	Gigler	Franz	Locomotive Fitter			
682	Krome	Erich	Locomotive Fitter		7.6.29	
683	Schnick	Walter	Locomotive Fitter		24.8.29	
684	Hellwig	Karl	Locomotive Driver		15.9.29	
685	Lorn	Friedrich	Dredger Operator		27.7.29	
686	Kratz	August	Excavation Foreman		23.6.29	
687	Danzck	Anton	Excavation Foreman		24.5.29	
688	Buchmann	Max	Locomotive Fitter			
689	Baumann	Karl	Locomotive Fitter		27.11.28	
690	Brahm	Robert	Locomotive Driver			
691	Kazschollek	Emil	Mechanic		24.11.28	
692	Zask	Hugo	Excavation Foreman		26.10.28	
693	Nauzlick	Franz	Locomotive Driver		12.4.29	
694	Hinzerheimer	Ludwig	Excavation Foreman		27.7.29	
695	Feistaner	Anton	Dredger Operator		31.10.28	
696	Engartner	Michael	Excavation Foreman		4.10.29	
697	Keller	Johannes	Mechanic			
698	Leu	Wilhelm	Locomotive Fitter			
699	Weise	Johann	Dredger Operator		17.3.29	
700	Wichofsky	Wilhelm	Carpenter			

No:	Surname	Christian Name	Occupation	Date of Arrival	Date of Return	Existing Permit Expires
701	Schatz	Wilhelm	Carpenter		31.10.29	
702	Goetze	Hermann	Mechanic (Turner)		30.6.29	
703	Keldenhauer	Willi			8.3.29	

No:	Surname	Christian Name	Occupation	Date of Arrival	Date of Return	Existing Permit Expires
704	Klump	Karl	Machinist for oxygen plant			
705	Dommerz	Adolf	Locomotive Driver		16.1.29	
706	Beautemps	Peter	Mechanic		15.8.29	
707	Radsey	Johann	Locomotive Driver		31.1.29	
708	Gruening	Gustav	Locomotive Driver		14.5.29	
709	Boeller	Karl	Reinforced Concrete Erection Foreman		24.8.29	
710	Healitreck	Heinrich	Rockdriller – Miner		14.11.28	
711	Vattes	Walter	Dredger Operator		2.5.29	
712	Krenys	Josef	Mechanic		16.9.28	
713	Schaletzki	Josef	Mechanic (Turner)		23.6.29	
714	Schroeder	Franz	Locomotive Driver		21.4.29	
715	Pogoda	Anton	Carpenter			
716	Gugel	Anton	Special Concrete Worker		24.8.29	
717	Schunk	Albert	Locomotive Driver		12.7.29	
718	Resschke	Erich	Locomotive Driver		11.1.29	
719	Sellau	Paul	Locomotive Driver		16.1.29	
720	Roeber	Karl	Mechanic		15.1.29	
721	Rampf	Otto	Dredger Operator		24.5.29	
722	Kuck	Georg	Rockdriller for Belgian Boring Machines		27.7.29	
723	Stiller	Peter	Rockdriller for Belgian Boring Machines		7.6.29	
724	Sperlich	Paul	Loco-Fitter			
725	Blaser	Joseph	Mechanic (Turner)		23.6.29	
726	Weingarten	August	Mechanic			
727	Amer	Georg	Rockdriller & Miner		30.11.28	
728	Kaffka	Johann	Special Concrete Worker		8.11.28	
729	Meyer	Gottlieb	Dredger Driver *(Returned to Ireland – his previous No. was 181)*		8.11.28	
730	Morlock	Matthaeus	Dredger Driver		24.8.29	
731	Bauerfeind	Oswin	Excavation Foreman			
732	Kistuch	Lee	Excavation Foreman		22.3.29	
733	Bockelberg	Adolf	Loco Driver		30.6.29	
734	Kemer	Johann	Carpenter		13.2.28	

No:	Surname	Christian Name	Occupation	Date of Arrival	Date of Return	Existing Permit Expires
735	Gommans	August	Loco Fitter		2.11.28	
736	Kinke (Lincke)	Obert (Robert)	Turner		13.2.28	
737	Dieterle	Christian	Mason		15.8.29	
738	Straubinger	Karl	Dredger Driver		7.11.28	
739	Greilinger	Ernst	Excavaton Foreman		27.7.29	

– 20 –

No:	Surname	Christian Name	Occupation	Date of Arrival	Date of Return	Existing Permit Expires
740	Breitfeld	Hans	Mechanic			
741	Paeperek	Rudolf	Mechanic			
742	Behrwind	Paul	Mechanic			
743	Weiher	Willy	Mechanic			
744	Kupper	August	Excavation Foreman			
745	Rose	Alfred	Mechanic			
746	Vleck	Josef	Special Concrete Worker		27.7.29	
747	Franke	Wilhelm	Mechanic		23.6.29	
748	Bartczak	Michael	Dredger Operator		6.4.29	
749	Schell	Josef	Mechanic		14.6.29	
750	Krell	Paul	Excavation Foreman		7.11.28	
751	Azchiernig	Johannes	Rockdriller		21.8.28	
752	Grasser	Joseph	Escavation Foreman			
753	Stahlschmidt	Wilhelm	Excavation Foreman		16.11.28	
754	Nenner	Kurt	Dredger Operator		28.8.29	
755	Straub	Joseph	Dredger Operator		13.2.28	
756	Seiler	Otto	Special Concrete Worker		21.9.28	
757	Schlachter	Alfred	Rockdriller		9.8.29	
758	Pohl	Wilhelm	Mechanic			
759	Pflugstert	Adolf	Mechanic			
760	Brau	Richard	Student		10.10.28	
761	Schmidt	Karl Friedrick	Mechanic			
762	Breschnig	Wilhelm	Dredger Operator			
763	Hellfritzsch	Edwin	Excavation Foreman		23.6.29	
764	Mensel	Bernhard	Fitter for stone crushing plant		30.6.29	
765	Cornelius	Karl	Excavation Foreman		27.7.29	
766	Corny	Wilhelm	Rockdriller		17.3.29	
767	Siebe	Paul	Foreman Fitter		22.3.29	
768	Lutz	Fritz	Student		15.11.28	
769	Berger	Joseph	Granite Worker			
770	Kaselmozer	Johann	Granite Worker			

No:	Surname	Christian Name	Occupation	Date of Arrival	Date of Return	Existing Permit Expires
771	Kniebe	Wilhelm	Electrician		9.8.29	
772	Wetzlich	Willy	Mechanic		9.8.29	
773	Boeser	Simon	Granite Worker		31.1.29	
774	Karkowsky	Paul	Mechanic		13.10.28	
775	Marsollek	Josef	Dredger Operator		7.7.29	
776	Schefold	Alphons	Mechanic		16.12.28	
777	Helmers	Johannes	Locomotive Driver		13.2.29	
778	Hartwig	Peter	Excavation Foreman		23.6.29	

No:	Surname	Christian Name	Occupation	Date of Arrival	Date of Return	Existing Permit Expires
779	Gross	Anton	Mechanic		10.5.29	
780	Wanner	Felix	Reinforced Concrete Creator		22.2.29	
781	Greilinger	Ludwig	Excavation Foreman		25.4.29	
782	Muench	Joseph	Cement Worker		31.1.29	
783	Merkel	Joseph	Granite Worker			
784	Eberl	Ludwig	Cement Worker		4.10.29	
785	Hainke	Walter	Mechanic for Cranes		31.1.29	
786	Ehlers	Hermann	Excavation Foreman		25.4.29	
787	Junger	Oskar	Reinforced Concrete Erector		31.10.28	
788	Steckel	Wilhelm	Reinforced Concrete Erector		25.4.29	
789	Schultz	Walter	Turner			
790	Nekardt	Heinrich	Locomotive Driver		7.6.29	
791	Findeiss	Johann	Electricial		13.2.28	
792	Geutsch	Wilhelm	Mechanic		30.6.29	
793	Dorn	Willy	Mechanic for cable crane		20.7.29	
794	Bernhold	Hans	Student		31.1.29	
795	Will	Willi	Mechanic for cable crane		20.7.29	
796	Koestner	Heinrich	Locomotive Fitter		16.10.29	
797	Steffen	Willi	Rockdriller		12.7.29	
798	Kupfer	Otto	Reinforced Concrete Worker		8.3.29	
799	Nielksoeg	Gustav	Excavation Foreman		27.7.29	
800	Steinberg	Karl	Reinforced Concrete Worker			
801	Luedemann	Wilhelm	Excavation Foreman		27.7.29	
802	Reihwald	Heinrich	Reinforced Concrete Worker		12.7.29	
803	Kuehne (prev.No.647)	Hermann	Mechanic		30.6.29	

804	Hotter	Rudolf	Rockdriller and Miner		1.6.29	
805	Schuett	Wilhelm	Swivel Crane Driver		20.7.29	
806	Geunrich	Karl	Swivel Crane Driver		20.7.29	
807	Casper	Georg	Copper smith		31.8.29	
808	Nodler	Albert	Excavation Foreman		24.5.29	
809	Naindl	Rudolf	Excavation Foreman		23.6.29	
810	Breuer	Joseph	Excavation Foreman		25.4.29	
811	Ringelstetter	Matthias	Excavation Foreman		25.4.29	
812	Thoms	Albert	Excavation Foreman		29.7.29	
813	Narnisch	Otto	Reinforced Concrete Worker		16.4.29	
814	Beur	Albert	Excavation Foreman			
815	Knuz	Friedrich	Plasterer		12.5.29	
816	Brutloff	Walter	Driver		6.4.29	
817	Gross	Georg	Copper-smith			

– 22 –

No:	Surname	Christian Name	Occupation	Date of Arrival	Date of Return	Existing Permit Expires
818	Stahlschmidt	Wilhelm	Excavators Foreman		24.5.29	
819	Obernuber	Franz	Plasterer			
820	Rajdamowicz	Michael	Plasterer		7.7.29	
821	Rinze	Komad	Plasterer			
822	Kurt	Chrestel	Student		24.4.29	
823	Steingreeber	Emil	Plasterer		9.8.29	
824	Koeiting	Paul	Student *xxxtransferred to Siemens Schuchertwerke*		9.8.29	
825	Hueser	Hartwig	Student		20.9.29	
826	Paustear	Adolf	Diver			
827	Homburg	Gustav	Joiner			
828	Kranz	Hermann	Specialist		9.8.29	
822-829	Albert	Walter	Student		20.9.29	
823-830	Raschig	Heinz	Student		11.10.29	

Appendix 4

Valuation of Shannon Fisheries

by Mr Arthur Taylor
Valuer for Department of Industry and Commerce, 17 October 1929

(1)　I have been asked to make a valuation of the Salmon and Eel Fisheries on the Shannon River. I have also been asked for my observations on the probable sum that might be payable if the whole of these Fisheries were acquired by the State, and the probable sum that might be payable by way of compensation in the event of some of these Fisheries being interfered with by the Shannon Works.

(2)　In making my valuation, two severe handicaps have been imposed upon me but I recognize that these were unavoidable. In the first place, ordinarily, I should be supplied with maps and schedule of particulars showing the location of the various Fisheries, the particulars of ownership, and particulars of how they were held and worked. As this information does not appear to have been available, I have had to obtain these particulars for myself.

(3)　In the second place, it was pointed out to me that my enquiry should be regarded as one of a confidential nature. In gathering the necessary information I have been hampered by this restriction. In the circumstances I have done the best I could but, naturally, I am reluctant to give a valuation figure until I should be able to verify my information. It has been pointed out to me, however, that it is urgent that my report should be made available at once, and, therefore, I submit it with its imperfections. I have gone closely into the matter, and shall be surprised if my figure of £195,000 for the purchase of the entire Fisheries from the source to the end of the Estuary of the Shannon is wide of the mark. As explained in the body of my report it differs from my interim rough estimate by £17,590.

(4)　It is difficult to make a comparison between the probable cost of acquisition and the probable cost of compensation. In the first place, I have not been supplied with any date as to the probable or possible interference with the Fisheries by the Shannon Works. On this matter I found the widest divergence of opinion. Some said that in consequence of the Shannon Works the salmon would desert the river. If this happened it is clear that the cost of acquisition and the cost of compensation would be equal. The fact, however, that large salmon have been seen to go up the new fish ladder and that some have forced their way under the weir gates and have gone upstream disposed of this opinion.

(5) Salmon are obstinate fish and always nose against the strongest current in the river. When they spawn, however, they come down to the sea tail first. This does not much matter because 95 per cent of the cock salmon die on or in the neighbourhood of the spawning beds, and a great number of the hen salmon, after spawning, also die. But the fry too come down to the sea tail first and many good judges fear that the descending fry may be killed in the turbines. After consulting engineers on the subject and others who have made a life study of salmon, I am of the opinion that this view may be discounted and that relatively few fry are likely to be destroyed by the turbines, and almost certainly not more than are destroyed at present annually either by being trapped in the Eel weirs and Lax Weir or caught by unskilled anglers who mistake them for trout.

(6) The third consideration is whether in the dry summer months there will be sufficient water left in the river to enable the salmon to run. This is at least debatable from all I can gather. It is certain, moreover, that in sum, all the water in the river below the weir at Parteen will be low. This would suit the Lax Weir and the Abbey Fisheries to a nicety. No fish could pass these obstacles. In consequence all the Fisheries at Castleconnell and above would in all probability become a memory in a few years. My experience is that the people of County Clare and Limerick require only the shadow of an excuse to make a substantial claim for compensation. To put it plainly these people and their friends and advisers are prepared "to swear a hole through a pot", as the saying goes, in support of their claim. I can, therefore, foresee a host of vexatious and troublesome claims in the years to come if the Fisheries are not acquired now, and I believe good cases for compensation could be drawn up.

(7) It follows that compensation would have to be paid for these. As the amount payable would be so substantial it would appear to me unwise to pay this large amount and get nothing in return, while for a relatively small additional outlay the whole industry could be purchased with a chance of recouping the purchase price, and, perhaps, earning a profit on it if properly managed. In this connection it is worthy of consideration that Mr Berrington, Chief Commissioner of Fisheries (England) valued the Shannon Fisheries at between £40,000 to £50,000 a year; (*vide* page 30, question 577 of appendix to Report of the Commissioners of the Irish Inland Fisheries Commission 1901). There are, of course, objections to State owned Industries, and I cannot remember one which returned a profit, but this is not a good reason why the Shannon Fisheries should not prove an exception if the lesson taught by other failures is taken to heart.

(8) In such an enterprise there is always the element of speculation, but on the facts before me I incline to the view that in the special circumstances of the Shannon it is only common sense to consider carefully the proposal to acquire all the Fisheries. Before a definite decision is reached I would recommend that the following gentlemen be invited to a conference to express their views: Mr Hickman, Solicitor, Ennis; Mr Blood-Smyth, Solicitor, Limerick; Mr Garvey, Solicitor, Ballina, and Mr S. Whipp, Fort Henry. These gentlemen have expert knowledge of salmon fishery, and are shrewd practical men, in a position to give valuable advice on the matter.

(9) I particularly desire that this report of mine shall be regarded as most confidential and that the figures given by me should not be disclosed to any one except to those who must see them. My reason for this request is that in the course of my enquiries I found that information concerning the acquisition proposal had leaked out and that certain parties were out to make hay while the sun shone. If it falls on my shoulders to negotiate with the fishery owners it would be fatal if they became aware of my valuations. There is a great deal more than meets the eye in negotiating compensation or purchase. If Mr MacMahon, who has assisted me in making my enquiries, is permitted to assist me in my negotiations (I am presuming acquisition) I entertain the hope of securing some of the fisheries below my valuation estimates.

(10) In arriving at my estimates I have had to take such information as I could gather and I have not always been able to verify it. The figures set forth below are derived from some figures ascertained and from a comparison between these and the relative merits of other similar fisheries. It is only right to add, however, that I am for the most part in the dark as to the actual values. No opportunity has been afforded me of examining the account books to ascertain the value of the catches of fish, the wages account and the cost of upkeep etc. These accounts should be inspected by me to enable me to put a proper valuation on the various fisheries. If legislation is contemplated, and I understand it is, this point should be borne in mind. It might even be considered whether it would not be desirable, to base the purchase price on the average profit over, say, the past five years as disclosed in the relevant Income Tax Returns. No doubt this would be a novelty but it would be a great safeguard against exaggerated accounts. My information is that accounts are being bolstered up. It should not apply to a Rod Fishery because its value to a sportsman is only regulated by his pocket. A gentleman who is addicted to Salmon Fishing will willingly give a rent for a good piece of water out of all proportion to the value of the fish caught in it.

(A)   Estimate of Value of Salmon Fisheries from Mouth of Shannon Estuary to Lax Weir Co. waters at Corbally

(a)   Stake Weirs

(11) The Stake Weirs are fixed engines for fishing. In appearance these are crudely constructed and in fact are the deadliest fishing tackle known. Stakes are fixed in the river bed and run out at varying lengths for a long distance from the shore generally at a right angle with a T piece at the end. Nets are attached to these stakes. There is supposed to be a gap, being a free passage 4 feet wide in the structure through which salmon might escape and get past. In practice, however, by various ingenious means, and also by accident the gap is frequently rendered ineffective. These fixed engines are supposed not to fish during the close season at each weekend. Actually, however, at least some of the weirs do not invariably observe this rule. Normally these fixed engines may be employed only at the incoming tide. An exception is to be noted in respect of the weirs at Clounderlow Bay. These, on account of the natural conditions, can fish not only at the incoming tide but also at the outgoing tide. These weirs are said to skim the cream of the fish from the river.

(12) Up to about 1860, the number of these stake weirs grew each year until a point was reached (probably the number was 126) when the salmon fisheries of the whole of the Shannon were menaced. In consequence a Government enquiry was set up by the British Government, and as a result the number of stake weirs was cut down by the Act of 1863, to those who were able to establish a right for some years prior to 1860. I mention this as it may be a useful precedent to bear in mind. After the passing of the 1863 Act, the number of Stake Weirs was limited to 43.

(13) It would not appear to serve any useful purpose to give the further history of these Stake Weirs. The following figures will be useful, however, for the purpose of comparison and to drive home a point which I propose to make:

Year	No. of Stake Weirs in use
1860	43
1894	41
1904	36
1914	37
1920	23
1921	14
1922	11
1923	5
1924	6
1925	9
1926	12
1927	16
1928	29
1929	32

(14) It will be seen that from 1920 to 1924 there was a steady decline due, in part, no doubt, to the disturbed conditions of the country politically. From 1925 to 1927, there was a gradual incline. In 1929, the number increased to 32, or more than six times the number in use in 1923, and double the number in use in 1927. I am satisfied that the increase is due largely to the fact that the proposal mooted to acquire the Shannon Fisheries leaked out and that the disused Stake Weirs were then revived simply for the purpose of extracting compensation from the Government.

(15) It is possible that if the weirs were not acquired by the Government and if Commissioners were appointed with powers to enforce the compulsory use of the free passage to make regulations for the proper use of the weirs as fixed fishing engines, and to increase the week end close seasons, when it would be illegal to fish the weirs, that in a short time the revived weirs would again go out of action.

(16) However, I have to take matters as I find them and have to estimate the value of the Stake Weirs as they stand. In doing this I draw attention to the handicap under which I labour and to which I refer in paragraph 10 of this report.

(17) The number of Stake Weirs in the Shannon Estuary for which Licences were taken out in 1929 is 32. Attached is a list of these (cf pp. 300ff. at end of this Report).

(18) In examining these tables the following notes will be useful:-

Table 1: It will be seen that in Table 1 some necessary particulars have not been obtained. To obtain them it would be necessary to make direct enquiry and it was considered desirable not to do this. It is probable that some of the unnoted weirs in Table 1 appear in Table 2 under another description. The tables are compiled from official lists but these appear to be out of date and the descriptions do not always correspond with the names by which the weirs are known locally.

> **a)** I am unable to trace Kilcolgan Lower Weir but William J. Hayes undoubtedly took out a Licence in respect of it for 1929. In many cases the named occupiers of the Weirs are only names of persons acting for other people. This subterfuge is adopted for obtaining additional votes in the Elections of the Board of Conservators. A like remark applies to names of Licensees of draft, drift and snap nets and also salmon licences. Behind these names are the personalities of Messrs McKibbin and Anthony Mackey. Very probably Mr Pegum is also in the alliance. No doubt there is a similar alliance, indeed I know there is on the other side. I mention the matter to indicate the difficulties of obtaining precise information on any matter relating to the Shannon Fisheries.

> **b)** I have made no enquiries of T.C.D.

> **c)** Mr Hickman's weirs are reputed to be the best weirs in the Estuary. In

1928, the value of the fish killed was £1,400. He pays £200 a year to who-letime fishermen.

d) The actual owner of this weir is Mr Anthony Mackey, Castleconnell.

e) These weirs probably are the same as Kilkerrin Weir shown in Table 2. I believe that the actual owners are the Lax Weir Co. or Mr Barber of that Company. See (h).

f) These weirs probably are the same as Burrane Weirs shown in Table 2. The fishery consists of 6 weirs. It appears to have been purchased 20 years ago in the Land Judges Court for £3,600 by Mr Barber of the Lax Weir Company, and was rented by him to McAuliffe at a rental of £110 per year. My information is that it has since been purchased by Mr McKibbin for £2,200.

g) Used to rent at £300 per annum. Very fine weirs. Not in use because of family history.

h) These weirs were attached to the Lord Annally Estate. They were purchased by Col W.W. White in the Land Judges Court about 20 years ago for £7,500. I have not ascertained if Miss McAuliffe is the owner in fee or a Lease holder. Probably her "ownership" is only nominal in any event.

i) The Long Rock Weir is a very good one and possibly equal to Mr Hickman's weir. It is jointly owned by the Knight of Glynn and Mr McKibbin. It is noteworthy that the valuation of this weir was doubled last year although no official connected with the fishery asked for a revaluation. I believe compensation was the object of the increased valuation.

j) This is known also as Mount Pleasant Weir. It is the largest weir in the river and expensive to keep. The weir has not been fished for many years but is capable of being fished. If fished it would render Barnahard Weir practically useless. The present owner appears to be Mrs Mary McAuliffe who purchased the fee simple for £450 but she may have since disposed of her interest.

k) At Kilkerrin there are one or two weirs not fished for years. These belong to the British Government and were vested in the Secretary for War I understand.

(19) Options have been taken for the purchase of the undermentioned weirs at twenty years purchase of the rents.

> Barnahard
> Aylevaroo (Ballymote)
> Carrowndotia
> Colmanstown
> Mountshannon

All of the Vandeleur Estate and Slievedooley on the Henn Estate. I advise that these options to be taken up. The price is £4,900. The present rents upon which the price is based are much below the actual value. The position is that fear of the unruly people in the locality makes an open market impossible at present. If the option is not taken up Mr McKibbin will purchase the weirs. I must add that although Colmanstown and Mountshannon are shown as in fee on the Rated lists my information is that they are the property of the Vandeleur Estate. Something similar may occur in investigating the titles of the other weir but I have given the information vouchsafed to me.

(20) I have gone into the figures so far as information is available and I estimate that the Stake Weirs in the Estuary should be purchased for a sum of £48,250 or in round figures should be purchased for a sum of £50,000. I have allowed nothing for the Weirs no longer in use.

(B)   Drift and Draft Net Fishermen in the Estuary

(21) The Shannon Estuary from Limerick to the sea is 60 miles long and varies in breadth from two to six miles but immediately above Limerick city the river narrows in parts to about half a mile. In the Limerick district no Drift Nets are permitted but Draft Nets are. On the Fergus Inlets, etc., Draft Nets are also employed. These Draft Nets are about 80 yards long and in all about 60 are employed. Below Limerick where the river widens out Drift Nets, probably about

70, are employed. These nets are about 250 yards long by two yards wide. Where they can be used no other type of net can be used. By anchorage these Drift Nets can be made fixed engines and there is little doubt that this is often done. Before the Act of 1853 no Drift Nets were used in the Shannon Estuary. After the passing of that Act, the Drift Nets were introduced and are said to have largely nullified the advantage gained by the limitation of the Stake Weirs.

(22) The following table gives a list of the net licences taken out over a number of years:

Year	Drift Nets	Draft Nets
1924	61	27
1925	64	42
1926	76	41
1927	84	35
1928	94	20

The nets are costly and the upkeep relatively high. As a rule it takes three men to work a Drift or a Draft Net. Some of the net fishermen have never done anything else and when the Shannon season is over they make their way to the deep sea fishing. Other of the fishermen are small holders and take out a licence if the run of fish is found to be good. It is probable that in all between 400 and 500 persons are dependant in whole or in part for a livelihood out of this fishing, but the number of fishermen engaged is very likely not one-fourth of this number. My information is that each fisherman engaged would in a good season earn an average of about £3 a week for the season. The greater number of these men are known as the Coonagh men and live in the Limerick district. They are subsidized by a fishmonger in William Street, Limerick, who buys what they catch. My information is that in effect he controls them because he finances them. They are poor men and I understand that this fishmonger comes to their aid in 'bad times'.

(23) These fishermen exercise a public right. To extinguish this would probably cause a good deal of trouble and would be bound to inflict real hardship on the men and those dependant on them.

(24) At the same time it would not appear a business proposition to extinguish the Stake Weirs and give the Drift and Draft Net Fishermen in the Estuary and tideway a free hand. It might be possible to prohibit the use of the Drift Net and allow only the use of the Draft Net. Alternatively, the Drift Net could be restricted to the number which could establish a right over a period of years to be determined.

(25) I have carefully considered the matter of compensation, but am unable to find a basis on which to build. Strictly, even if the public right which these men exercise, was extinguished, they would not be entitled to compensation as they have no legal rights real or presumed. I have two suggestions to make for a satisfactory solution to the problem. One: Abolish the public right and in lieu, give the men affected undivided land in the possession of the Land Commission and a sum of money for stock. The land hunger is so keen in this area that the men will probably jump at this offer; Two: Restrict the right to take out licenses to those who have taken out licenses over a period of years to be determined, and let this operate only for the lifetime of the men who can now establish this right. Coupled with this could be a restriction as to the number of days on which the nets could be used.

(26) In any event there will be trouble most likely and poaching will probably increase. A Drift Net can be seen a mile off and, if it was liable to confiscation on being found to be used illegally, in a short time the illegal use of these nets would cease.

The Lax Weir Fishery

(27) Above the fishing ground fished by the Coonagh men, the water up to the County Boundary above Corbally House is claimed by the Lax Weir Company. The fishing is done, however, below the weir, but principally at the weir itself which extends across the river, nearly opposite Corbally House. In this weir is a King's Gap about 4 feet wide through which salmon can pass up the river.

(28) This Weir belongs to ancient history. It consists of two parts: one, the better part, was built of cut stone some one thousand years ago by Monks; the other part, not so substantial, but no less effective, was built subsequently, and in the 1916/22 political trouble was levelled and subsequently rebuilt. By charter, it was granted by King John to the Limerick Corporation which used to rent it. Subsequently the Corporation sold the fee simple to a Mr A. Bannatyne, who used to let it at £300 a year. After further vicissitudes the lease was purchased under favourable circumstances and subsequently the fee simple by a London Company of fish salesmen for about £20,000. This company was called the Lax Weir Company Ltd. From the annual return of 21.5.1923, it appears the number of shares was 9,000 ordinary £1 shares and 3,000 Debentures. The shares were issued as follows:-

H. W. Barber	2,997 shares
R. W. May	1,400 shares
W. H. Grant	1,500 shares
Forbes Steuart & Co.	2,998 shares
W. F. May	100 shares
Shares not issued	5 shares
	9,000

At present the shareholders are:

Henry Barber & Son Ltd.
William Richard Grant
William Forbes Steuart & Co.

Mr Stephenson represents William Forbes Steuart & Co. The Secretary is Mr W. O. Tottingham, Lax Weir Fishery Co., 33 Monument Street, London. The Manager of the Weir is Mr Liam Forde, Corbally, Limerick.

(29) For 1927 and 1928 the wages paid weekly varied from £110 to £130 each week, and for these two years the total expenses averaged £5,200 each year. I was unable to obtain definite information but ascertained that the net profit earned by this Company averaged over the past two years approximately £1,875 each year. I should like, however, to verify this figure. My information is that the profits widely fluctuated for various reasons and probably no profits were earned for several years. The principal reason for this was that the weir was badly mismanaged and the waters poached quite openly. It may be that malversation also occurred. When Mr Liam Forde was appointed Manager, the business immediately revived and in the past few years the profits have increased steadily. Last year was the best year for many years back. I have interviewed Mr Liam Forde and formed the impression, which other more immediately in touch with him have confirmed, that he is an exceptionally able and strong man. If the acquisition scheme goes through, I strongly advise that Mr Forde's services should be secured if possible.

(30) I have had very great difficulty indeed in obtaining the information given above. When on a private visit to London in May last, I took the opportunity of calling upon the Company. I had a long interview with Mr Tottingham, Secretary, but was unable to extract any valuable information. Subsequently, Mr Stephenson of the Lax Weir Co., called upon me in Dublin and I had a long interview with him. Later, at his invitation, I interviewed him in London at a Conference consisting of himself, Mr Grant and Mr Tottingham. Unfortunately, Mr Barber was unable to keep the appointment at the last moment. He appears to be the ruling factor, and obviously Mr Stephenson spoke from a memorandum supplied by Mr Barber. The conference lasted some hours. I am bound to say that it was only the tact and ability shown by Mr V. MacMahon, who accompanied me, that the barriers of restraint set up deliberately by these cold clear headed businessmen was broken down and that I was able to wrestle with figures and facts. In the long run cordial relations were established. Eventually, the demand was brought down to £30,000 for the fee simple of the Lax Weir Co.'s fishing rights and also Corbally House and premises, which go with it. This house is at present let at £100 free of rates to a Mr Quinn. The lease of the house is subject to a bead rent, but I did not take notice of the amount. In fact I took no notes at the interview because I explained that I regarded the whole matter as confidential and explained also that I was simply making enquiries with a view to valuation in the event of compensation being claimed, or in the event of it being found necessary to acquire the fishery. And I made it plain that it should not be assumed that either compensation might be payable or that the Fishery might be acquired. When the demand was got down to £30,000, I said that if I was buying the property, I would consider the amount a trifle high, and although I may be wrong, I am still of the opinion that my estimate of £25,000 is the probable figure that will be payable for this property in the event of acquisition. The rateable value of the property is £400.

(31) Perhaps it is well to mention here that both Mr Barber and Mr Stephenson are interested in the Stake Weirs in the Shannon estuary worked by Miss McAuliffe and Mrs Mary McAuliffe. So much was admitted as a result of my enquiry but it was vaguely suggested that they were not the owners. There is undoubtedly a business arrangement between the parties.

The Abbey Fishermen

(32) There is no finer fishery in the Shannon than that owned by the Abbey fishermen. For the most part they are illiterate, rough type of men, who live on the outskirts of Limerick city. They are an extremely jealous lot and will not allow any outsider in. They really form a distinct clan. They boast they are the oldest fishermen in Ireland and their history goes back over 500 years. Although only 13 crews of fishermen (4 men forming a crew) are employed in the fishery, the number dependant upon this fishery is probably 300. This is maximum and the figures obtained by me vary from 10 crews of 4 men each crew, 11 crews of 4 men each, to 14 crews of 4 men each. Very likely 13 crews of 4 men each is correct.

(33) These fishermen are not very particular as to how they fish or on whose waters but normally they fish by means of a snap net. For many years I have known that the Abbey men have an uncanny instinct which tells them where a fish is and their skill is so great that no fish can elude their determination to catch it.

(34) The representative of these fishermen is a fishmonger in Limerick named Mr Denis Hayes, Roches Street. I am told that the average earnings of these men varies in a good season from £156 to £260, or an average for each man from £3 to £5 per week. In a bad season the earnings probably drop to 10/- a week per man. I should say that deductions should be made for cost of upkeep of boots and nets and interest on capital invested; but within the narrow limits of my inquiry it was not possible to probe into exact details.

(35) The fishing season lasts 2½ weeks from 12th February to 12th July each year, but probably the state of the waters does not permit fishing for more than an average of 16 weeks in each season.

The extent of the fishery so far as I could as certain is as follows:

(36) From Corbally Mill Dam on Limerick shore to boundaries of Prospect Demesne. The Abbey men purchased this stretch from Col. Eyre Powell for £1,000 a few years ago (and, a matter of interest Col. Powell's ancestors got it as a grant from King Charles). On the Clare shore the Abbey men fish from the Mouth of the old Blackwater river to the boundary of Garraun townland. They purchased the fishing rights from the adjoining farmers on the stretch about 10 years ago for £180. They also fish from Garraun to Illaunareum Trench. This is the property of Captain Arthur who rents it to the Abbey men for £6.5.0. per annum. From Illaunareum Trench to Egan's boundary is the property of Pk. Moloney who rents it to the Abbey men for £3 a year. The Abbey men also own the stretch adjoining Egan's and Holmes' lands whose rights they purchased for £400. I have not been able to ascertain definitely the rent paid by the Abbey men to Mr D. R. O'Brien for his waters; but the rentals paid do not include Rod fishing. If the nets were taken off the Abbey men's fishing ground it would be a valuable rod fishing property and could be let at substantial rents.

(37) It is difficult to say the amount of compensation which would secure peaceful occupation of the waters fished by the turbulent Abbey men. The figure £40,000 has been mentioned to me. My second estimate is £26,000 as against £20,000, my original estimate.

(38) I have not ascertained the rateable value of this fishery. It does not appear to be rated as a distinct fishery but rates for part of it are shown in the Table annexed to the Rod Fishing section of my report which follows.

Rod Fisheries above the Abbey fisherman's property and below the Killaloe Fisheries

(39) (a) Major Ingham's Fishery as well as those which follow, unless otherwise stated, is situated on the Limerick shore of the Shannon River. It extends from O'Briensbridge down to Woodlands Fishery. It is the property of Major Ingham who purchased it about 1893 in the Land Estate Courts for £5,000. Rateable Value £110.

(b) Woodlands Fishery Above this is the Rod Fishery owned by Mr Malcolm D. Shaw and known as Woodlands Fishery. It is held under Fee Farm

Grant from Mr T. Johnston Stoney, Castleconnell, and is, no doubt, subject to an annual payment under the Fee Farm Grant but I have not so far ascertained the amount. The Fee Farm Grant includes the lands (about 40 acres) and the house. Mr Shaw's predecessors paid £5,000 for the property, but Mr Shaw purchased the property for something over £2,000 but less than £3,000. Rateable value £71.

(c) **Hermitage Fishery.** This fishery rivals Doonass Fishery mentioned below. It was purchased two years ago at public auction for £3,080 plus fees and costs. It is not the property of Mr S. Whipp, Fort Henry, who has let it at a rental of £200 per annum.

(d) **Prospect Fishery** (Clareville). Formerly this was a separate Fishery and an excellent one. It has now been divided into two parts (a) Clareville which is about 500 yards long is owned by Georgina Powell, Mansfield, Passage West, Cork; David Hall, Dromona, Co. Waterford; and Evelyn H. Johnston. These appear to be the reps of Colonel Eyre Powell and Johnston, deceased, the former owners. My information is that it is let at a rental of £175 a year to a Captain Foster on a seven year lease, with power to break it in 1930. The water in the lease includes the second part of Prospect (b) which is North of Clareville and is held in fee by Mr J. F. E. Morley. Rateable Value of this Fishery appears to be £87.10.0.

(e) **Newgarden Fishery.** At one time this was a magnificent fishery and used to let at £500 per annum. It is held in Fee Simple by Mr J. F. E. Morley. He has let it at present for £250 a year on lease jointly to Lord Kingstown and Mr George White. My information is that the term of the lease is 7 years and that the lease includes a clause enabling it to be broken in 1930. Rateable Value £50.

(f) **Ernest Brown's Fishery.** This Fishery does not appear to be at present of much value. This arises from two causes (1) that it is continually poached by the Abbey fishermen, (2) the water is so high normally that it is not fit for fishing except an uncertain period when the water is right for fishing.

Clare Side of River Rod Fisheries

(g) **John Meaney's Fishery** (Belle Isle House). This Fishery is opposite

Ernest Brown's on the Limerick side. In value it is similar to Mr Brown's Fishery and my report on it applies.

(**h**) Above this – going towards Killaloe – is Matthew Egan's Fishery. It is sandwiched in between Mr Meaney's Fishery and Landscape Fishery. My information is that it is in the same boat as Mr Meaney's and Mr Brown's Fishery mentioned above.

(**i**) **Landscape Fishery** comes next. It is the property of Mr Hugh O'Brien Moran, Solicitor and County Registrar. For the 1929 season he let this Fishery for £140 for the season. The Fishing is an appurtenance of the house and lands adjoining which Mr O'Brien Moran purchased three years ago for £3,500 plus fees and costs. The Rateable Value of the Fishery is £70.

(**j**) **Doonass Fishery** follows. This is reputed to the best fishery on the Shannon. It was sold over a year ago by public auction and fetched £2,200 plus fees and costs. The purchaser was Col. Roche Kelly, Islandrum. The Rateable Value of this Fishery is £150.

(**k**) **Summer Hill Fishery**. It is reputed to be a very good Fishery. It was the property of Col. Arthur Vincent (Deceased) and is now held jointly by five Minors, his representatives. It is let to Major Ingham who has fished it for years. The rental payable by Major Ingham is probably £100 a year or perhaps a little more. The Rateable Value is £100 a year.

(**l**) **Waterpark Fishery**. This is the property of Mr John Hartigan. My information is that it is of little value. On the other hand it will surprise me greatly if Mr Hartigan won't make a big demand for it.

(**m**) **Rose Hill Fishery**. This is below Mr Hartigan's. The tenant of the Fishery and the purchasing tenant of the adjoining lands under the 1923 Land Act is Mr Anthony Mackey, Castleconnell. My information is that this Fishery is of no value. Nevertheless, in the letter dated (not given) from the Secretary, Irish Land Commission to you it is stated that it was proposed to vest this Fishery in Mr Mackey and I understood the amount apportioned in respect of it is £200. You will recollect I wrote to you about this Fishery on the 23rd January 1929, and subsequently interviewed you when the Land Commission proposal was received. The Rateable Value of the Fishery appears to be £10. As the Fishery is vested in the Land Commission it appears to me it is the Minister who should take steps to acquire it.

(n) **Mr O'Shea's Fishery** is next. I know nothing about it except that it is reputed to be of little value, as it can be fished only at uncertain periods in consequence of the water being too high. From this to O'Briensbridge the water appears to be free.

(o) In addition to the above I have extracted the following particulars from the Register kept by Mr Alton, Secretary of the Limerick Fishery District. I am not satisfied that these particulars are correct.

COUNTY CLARE

Electoral Division	Townland	Ord. Sheet	Occupier	Immediate Lessor	Rateable Value
Ballyglass	Clarefield	63	Thos. Cloney & Partner	In Fee	£5. Probably Lax Weir own this
Cappavilla	-	53/54 63/4	do.	Arthur Minors	£5
Keltenanlea	-	53/4 64	Pk. Burke	In Fee	£3
Do	-	53/4	Mrs C Burke	do	£7
Do	-	53/4 64	Pk. O'Meara	do	£1.10.0.
Do	-	53/4	A. Mackey	N.A.Robins Book above	£10 See C
O'Brien'sbridge	-	54	Miss McCraith	In Fee	£3
Killaloe	-	37/45	S. Whipp	Gen. O'Niall	£10
Do	-	45	Do	BOW	£15

Except Mrs C. Burke's which is similar to A.Mackey's & J. Hartigan's Fishery I am unable to identify the above Fisheries. The list down to Mr Mackey may refer to Fisheries appurtenant to lands purchased under the Land Acts, but none are important Fisheries apparently. The figures for Whipp appear to relate to his Fort Henry Fishery and it will be easy to verify this by inquiry later on.

Killaloe Fisheries

(41) Fort Henry Fisheries. This was a very fine Fishery and the property in fee of Mr S. Whipp, and is fully described in my report on the compensation agreed viz. £9,200. In the circumstances it is not necessary to say anything more about it here. Rateable Value £30.

(42) Major Lefroy's Fishery. The rateable value of this Fishery is £10 and the rental paid for it by Mr S. Whipp is £80 per annum. It is known as the Pool Fishery, is situated opposite the Mills at Killaloe, and is the property of Major Lefroy. The term of Mr Whipp's lease of this Fishery is 15 years from 1st February 1924. There is a clause enabling the lease to be broken should the Shannon Works interfere with the Fishery. This has come to pass. Formerly the Fishery was rented once at £120 and also, more frequently, at £100 a year. In addition to the Pool Fishery, Major Lefroy holds a reversionary interest in another Fishery adjoining, which is let to Mr Frank Whipp at a rental of £1 per year payable to Major Lefroy.

(43) The Bishop's Fishery. This Fishery extends from below Major Lefroy's Fishery at Killaloe to Clarisforde. It was formerly attached to the Bishopric. It was purchased in 1920 by Mr Frank Whipp, Rockdale, the present owner from General Spaight. (Mr Frank Whipp owns only portion of 'the Bishop's Fishery' as the other portion was purchased by Mr Sam Whipp, his brother.) Although held in fee simple, the purchase was conditional on the payment by Mr Frank Whipp of annuity of £35 per annum to an old lady during her lifetime. Mr F. Whipp fished for portion of the year, and let it each year for May and three weeks in June for £120 plus £5 wages each week to two men whom he employs all the year round.

(44) Mr John Crowe's Fishery. This is the property of Mr John Crowe, Draper, Killaloe and Dundrum, (Co. Tipp.). When buying his business premises at Killaloe he had to purchase the Fishery also as a condition of sale. He let the Fishery to Mr S. Shipp at £4 per annum. Rateable Value £22. Above this Fishery is free water.

(45) Reps of Roger O'Farrell, Cloonfadda. This Fishery is the property of Reps of Roger O'Farrell (decd.). It was formerly the property of Lord Leconfield who used to rent it one time at £20 a year to Mr Denis Crowe of Crowe's Lough. When Lord Leconfield sold his estate to the tenants, he did not reserve the sporting rights, and these became vested in the tenants. Denis Crowe, however, continued to fish it, and although he had no claim to the Fishery he was the rated occupier and paid the rate. Recently in Killaloe Courts, District Justice Gleeson decided that Mr Crowe had no rights in this Fishery. It is the property of Reps. Of Roger O'Farrell, the whole of whose lands have been acquired under the Shannon Electricity Act 1925. Rated Value £1.

(46) Mrs Kate Mulcahy, Cloonfadda. This adjoins the above and was rented by Mr S. Whipp from Mrs Mulcahy at a rental of £7 per year.

(47) Killaloe Net Fishermen. The position with regard to these is obscure. Formerly there were three crews of men, and in recent years only two crews. Claims may be anticipated from Mr Denis Crowe, Mr William Barry, and Mr Joseph Ahern (O'Brien's Bridge). So far as I have been able to ascertain these crews had no legal rights. They fished in free waters or by power of fear in other waters, and probably were not above poaching in preserved waters when opportunity offered. They may, however, have some legal rights not disclosed in some waters. At Killaloe also, there is free water which was fished by some fishermen and a claim may arise in connection with these. The parties, on the information I have been able to obtain, would not appear, however, to have any legal right to compensation.

Fisheries above Lough Derg

(48) Free Waters. The water at Portumna is free. Only a couple of Fords of little value are fishable. A like remark applies to Meelick and Banagher. Between Banagher and Shannon Bridge is Derryholmes also free and of little value.

(49) Suck Fishery. Shannon Bridge at the Mouth of the Suck, is free water.

Portion of the river Suck between Mouth and Ballinasloe is preserved by Mr Potts of Correen Castle. He catches Peal or Grilse, but as these run fast the number of fish caught is small and the fishing is confined to the month of June. A few salmon are caught at Ballinasloe when they run up with the Peal.

(50) Clonmacnoise. A small stretch of water near Bunthulla Hill is owned by Major Charleton who used to preserve it. He left the country some years ago. When the water is in good order for fishing, Athlone anglers resort there and catch a few salmon. It is treated as free water.

(51) Between Clonmacnoise and Athlone is County Pile and Wrens Island. This is free water.

(52) Athlone. Below the Weir at Athlone on County Westmeath side two well-to-do farmers named James and John Macken fish by means of draft nets. They don't employ anyone. The foreshore is rented by these men from the Board of Works for a small rent. They use two nets. Without a footing on the foreshore they could not work these nets. No Rateable Value.

(53) Lough Rea. Free rough fishing.

(54) Tarmonbarry. Free water.

(55) Boyle. There is a good fishery here owned by Sir Thomas Stafford.

(56) Longford. I have been unable to obtain particulars but understand that in this area the waters owned by Sir Charles Gunning who appears to have rented the waters to Mr Robert Farrell. Rateable Value £6.

(57) It will be observed that above Killaloe it may be presumed that there is no commercial fishery. With regard to these fisheries, I append the following information which I obtained from Mr Alton, but it is not very reliable as in case in which I have been able to verify I found the information out of date.

COUNTY WESTMEATH

Electoral Division	Townland	Ord. Sheet	Occupier	Immediate Lessor	Rateable Value
Athlone E.R.	Carrickobrien	29	John Galvin	In Fee	Nil (no value)
Athlone E.R.	Golden Island (Kilmaine)	29	Ml. Curley	In Fee	Nil (no value)
Athlone E.R.	Golden Island (Kilmaine)	29	Lord Kilmaine	In Fee	Nil (no value)
Athlone West	Athlone and Brigmeadow	52	Pat Finneran	J. Macken	Nil (no value)

(58) The tentative total purchase money for all the fisheries set out above in pars. 38 to 56, estimated by me was £53,420, but on the additional information before me I consider the amount should be increased to £59,000.

Eel Fisheries

(59) The Eel Fisheries on the Shannon are the property of the Office of Public Works. The Commissioners let these Fisheries to Mr Anthony Mackey of Castleconnell at a rental of £1,800 per annum. There are Eel Weirs at Meelick, Athlone, (these are practically derelict), Killaloe and Castleconnell. At various places, but principally at Killaloe and neighbourhood, Mr Mackey has established Eel Tanks in which the eels are stored. The Shannon Works have now submerged these and also the Killaloe Eel Weir.

(60) The eels begin to run to the sea in July but the bulk of the eels do not run until a dark stormy October or November night. Eel fishing always takes place at night. Eels will not run in moonlight. I mention this because an eel run is an extraordinary thing. As I say, the bulk of eels run on one or two nights in tangled masses. It is unlikely, therefore, that Mr Mackey will lose any eels this year. True, he cannot work his Killaloe weirs, and he states this is his best weir, but the probability is that he may catch the bulk of the eels at his weir at Castleconnell. On the other hand, the disappearance of the Killaloe Weir can be made out a distinct loss and may benefit the Lax Weir Co. by increasing their percentage of catch at Mr Mackey's expense. Mr Mackey's claim is for £15,000 compensation. I estimate that he may suffer, if at all, and this should be capable of proof from his accounts, to the extent of perhaps £5,000 at the outside. To the Board of Works the compensation payable should be in the neighbourhood of £28,000. My total estimate for the eel fisheries is £33,800 or in round figures £34,000. In estimating Mr Mackey's compensation I am considering not only the loss of eels but also the loss of property, and I include the Athlone and Meelick Weirs.

(61) Summary of Valuation

Stake Weirs in Estuary	£50,000
Lax Weir at Corbally	£25,000
Abbey Fishery	£26,000
Rod and other Fisheries, above Abbey Fishery	£60,000
Eel Fisheries	£34,000
TOTAL :	£ 195,000

17 October 1929

Table 1: Stake Weirs for which Licences Were Taken Out in 1929

Name of Weir	Townland	Licensee or Occupier	Immediate Lessors	Rate-able Value	Annual Rent	Lease Expires	North N or South S Side of River	Remarks
Barnahard	Carrownacalla S	Miss Burke	H.S.Vandeleur	15	32	1932	N 143	Married name Mrs B.Hennessy
Colmanstown	Colmanstown	Pk. Cullinane	H.S.Vandeleur	10	40	1932	N 106	Formerly Pk. McAuliffe. Real owner Mr McKibbin
Querrin Net	Shangarragh	Wm T.Counihan W.J. Hayes	In fee				N 132	See note (f) report
Kilcolgan Lr.				1			?	See note (a) report
Cloonaman No.2	Killeton & Cloonman	C.V.Hickie and D.O'Sullivan	Provost T.C.D	120			S 123	See note (b) report
Mountshannon	Mountshannon E.	James Healy	H.S.Vandeleur	1	30	1932	N 36	Formerly Pk. McAuliffe. Real owner Mr McKibbin
Slievedooley	Slievedooley	James Healy	F.B. Henn	30	40	1931	N 135	Formerly Pk. McAuliffe. Real owner Mr McKibbin
Kilmore Church	Kilmore	F.W.Gore Hickman	In fee	50			N 48	See note (c) report
Knock Net	Knock	F.W.Gore Hickman	In fee	30			N 139	See note (c) report
Coolnanoonagh		William Krepps					S 58 & 137	This weir really belongs to Anthony M Note (d) Probably 2 weirs
Tarbert Net	Tarbert	Major C.R.Leslie	In fee	45			S 87	
Carrowdotia East Carrowdotia West	Carrowdotia S	John Malone	H.S.Vandeleur	20	43	1932	N 144	No licence taken out in 1928. Formerly Pk.McAuliffe. Only weir rated & called Carrowdotia S. There are 2 weirs one known locally as Moyne. Real owner McKibbin

Place	Townland	Holder	Tenure	No.	No.	Year	Ref	Notes
Durnish Weir	Durnish	Lord Monteagle	In fee	30			S 56	
Mount Trenchard	Mt. Trenchard	Lord Monteagle	In fee	15			S 57	
Lacknabahee		Miss McAuliffe	See Table 2				?N 38	See notes (l) & (h) in report
Lackyle		Miss McAuliffe	See Table 2				N 39	See notes (l) & (h) in report
Millpark		Miss McAuliffe	See Table 2	20	60	1932	N 104	See notes (l) & (h) in report
Aylevarroo	Ballymote W.	Mrs Mary McAuliffe	H.S.Vandeleur	75 65			N 40 N 141	See notes (b) in report
Lynch's Point Rusheen Point		Ditto. F.McKibbin	See Table 2	60	110	1933	N 124	Not fished for years. No licence taken out in 1928. Rated occupier Marcus Keane. See (g) report
Scattery Is. D) Scattery Is.E)	Scattery Island	L. McAuliffe F. McAuliffe	In fee					
Long Rock	? Cahevagh	F.McKibbin	In fee	200			S 44	See note (l)
Poulnadavee	Poulnadavee	F.McKibbin	In fee	5			N 43	Formerly Pk. McAuliffe
Poulmagulky		F.McKibbin					? S	
Shannon Lawn		F.McKibbin					S 17	
Woodpoint		F.McKibbin		30			? S	
Carrigane	Carrigane	T.E. Pegum	In fee	15			S 52	
Carrowbane Beg	Carrowbane Beg	T.E. Pegum	In fee	25			? S 53	
Kyleatallin	Kyleatallin	T.E. Pegum	In fee	6			? S 49	
Kilpadogue No.1	Kilpadogue	T.E. Pegum	Major C.R.Leslie				? S 122	

Para. 17(b)   (2) STAKE WEIRS SHOWN AS RATED, BUT IN RESPECT OF WHICH NO LICENCE TAKEN OUT

Townland or Name of Weir	Licensee or Occupier	Immediate Lessors	Rate-able Value	Annual Rent	Lease Expires	North N or South S South of River	Remarks
Kilkerrion	Mary & Anne McAuliffe	In fee	60				See note (e) and Table 1 Lackyle, etc.
Burrane Lr.	Mrs Mary McAuliffe	H.W.Barbour	75				See note (f) and Table (1) Lynch's Point etc.
Burrane Lr.	Mrs Mary McAuliffe	H.W.Barbour (now F.McKibbin)	65				See note (f) and Table (1) Lynch's Point etc.
Clarefield	E.J.V.Ronaldson	In fee	Nil			N 107	Not fished for years – see note (j)
Ballyhoolahan	Knight of Glen	Reps. J.S.O'Leary	1				See note (i)
Foynes Island	Lord Monteagle	In fee	5				

Appendix 5

Engineers with Shannon Power Development

On 2 May 1928 Paddy McGilligan, Minister for Industry and Commerce, in reply to a Dáil question, stated that as of 1 May there were 39 Irish engineers employed on the Shannon Scheme. At the same date in 1927 there were 34. In the intervening period 28 had resigned – 13 then took up appointments at home and six went abroad; the reason for the resignation of the nine was not known.

The number of resignations seems extraordinarily high by any standards, but no reasons are given. These departures were during a time of peak development on the Scheme and must have created tremendous difficulties in continuity and in the execution of various elements of the project.

In April 1929 Professor Rishworth, Chief Civil Engineer with Shannon Power Development, was requested by the Department of Industry and Commerce to submit a list of applicants for appointment under the Electricity Supply Board. He did so and his communication ends with the note, 'There may be others worthy of consideration.' This would indicate that his list is not comprehensive, but he does enumerate 35 applicants, not including himself and Prendergast, Resident Engineer at the Strand Barracks. The list is:

1.   J. MacDonald
2.   J. MacLachlan
3.   W.I. Bloomer
4.   J.A. Sullivan
5.   E. Roche
6.   A.B. Killeen
7.   C. Stenson
8.   M. Keehan
9.   L.A. Jones
10.   A.St J. Kennedy
11.   R.P. Devlin
12.   E.V. Switzer
13.   W.T. McHugh
14.   V.D. Harty
15.   M.G. Ahern

16. J.H. Whitton
17. H. Hawe
18. J.J. Breen
19. J.A. O'Riordan
20. G.R. Buckley – Draughtsman
21. F. Crowley   – Inspector
22. W. Byrne   "
23. W.J. Chadwick   "
24. W. Fitzsimmons   "
25. V. O'Reilly   "
26. J. McCluskey   "
27. T. Nagle   "
28. St J. Harold   "
29. T.M. Randles   "
30. J.B. Cahill
31. J. O'Driscoll   "
32. J. Dore   "
33. M. Flannery   "
34. T. Nolan   "
35. J.J. Walsh – Siemens' Bauunion Staff.

Elsewhere in Professor Rishworth's papers (NA, SS14065) there is a list of Shannon Power Development engineers dating from March 1930, which gives their dates of appointment and salaries for most of them:

NAME	DATE	SALARY
1. J.K. Prendergast	?	?
2. J. MacDonald	22 June 1025	?
3. J. MacLachlan	10 August 1926	£600
4. W.I. Bloomer	5 October 1025	£500
5. J.A. Sullivan	14 September 1025	£500
6. A.B. Killeen	4 January 1927	£400
7. C. Stenson	23 May 1927	£400
8. E. Roche	October 1925	£400
9. M. Keehan	18 September 1925	£375-400

10.	A. St J. Kennedy	16 November 1925	£375-400
11.	L.A. Jones	1 February 1926	£375-400
12.	R.P. Devlin	8 November 1926	£300
13.	E.V. Switzer	8 November 1926	£300
14.	W.T. McHugh	8 November 1926	£300
15.	J.A. O'Riordan	13 February 1928	£300
16.	W. Hawe	1 October 1928	£250
17.	J.H. Whitton	1 October 1928	£250
18.	M.G. Ahern	4 October 1928	£250
19.	V.D. Harty	8 October 1928	£250
20.	J.J. Breen	22 October 1928	£250

The following were employed as Inspectors or Draughtsmen:

	NAME	DATE	SALARY
21.	V. O'Reilly	May 1927	£5 10s.
22.	W. Fitzsimmons	10 May 1927	£5 15s.
23.	J. Nolan	August 1928	£5 10s.
24.	J. Lane	10 June 1929	£4
25.	J. Burke	- - -	£4
26.	G.R. Buckley	May 1925	£5
27.	K. Crowley	27 April 1926	£6
28.	W. Byrne	August 1926	£6
29.	M. Flannery	July 1926	£5 15s.
30.	J.M. Randles	6 September 1926	£5 15s.
31.	J. Dore	14 may 1926	£5 10s.
32.	J. O'Driscoll	15 December 1929	£5 10s.
33.	L. Nagle	December 1929	£5 10s.
34.	J. Harold	21 May 1928	£5 10s.
35.	J. O'Driscoll	15 December 1926	£5 10s.

Index